Design Engineer's
Case Studies and Examples

Design Engineer's
Case Studies and Examples

Keith L. Richards

CRC Press
Taylor & Francis Group
Boca Raton London New York

CRC Press is an imprint of the
Taylor & Francis Group, an **informa** business

CRC Press
Taylor & Francis Group
6000 Broken Sound Parkway NW, Suite 300
Boca Raton, FL 33487-2742

© 2014 by Taylor & Francis Group, LLC
CRC Press is an imprint of Taylor & Francis Group, an Informa business

No claim to original U.S. Government works

Printed on acid-free paper
Version Date: 20130819

International Standard Book Number-13: 978-1-4665-9280-3 (Paperback)

Library of Congress Cataloging-in-Publication Data

Richards, Keith L.
   Design engineer's case studies and examples / Keith L. Richards.
      pages cm
   Includes bibliographical references and index.
   ISBN 978-1-4665-9280-3 (pbk.)
   1. Mechanical engineering--Problems, exercises, etc. 2. Machine design--Problems, exercises, etc.
I. Title.

TJ153.R48 2014
604.2--dc23   2013032503

Visit the Taylor & Francis Web site at
http://www.taylorandfrancis.com

and the CRC Press Web site at
http://www.crcpress.com

Contents

Preface...ix
About the Author xi

Chapter 1 Introduction to Stress and Strain..1

    1.1   Direct Stress 1
    1.2   Tensile Stress 1
    1.3   Compressive Stress..1
    1.4   Direct Strains..2
    1.5   Modulus of Elasticity (E) 3
    1.6   Ultimate Tensile Stress...4
    1.7   Shear Stress 4
    1.8   Shear Strain 6
    1.9   Modulus of Rigidity 6
    1.10   Ultimate Shear Stress 7
    1.11   Double Shear 8
    1.12   Poisson's Ratio 9
    1.13   Converting between Stresses and Strains.................................... 10
    1.14   Three Dimensional Stress and Strain 11
    1.15   Volumetric Strain 11
    1.16   Bulk Modulus 12
    1.17   Relationship between the Elastic Constants................................. 12
    1.18   Factor of Safety in Tensile or Compressive Mode....................... 13
    1.19   Factor of Safety in Shear Mode 14
    1.20   Theories of Elastic Failure 14
           1.20.1   Rankine's Principal Stress Theory................................ 16
           1.20.2   St. Venant's Maximum Principal Strain Theory.............. 16
           1.20.3   Shear Strain Energy Theory (Von Mises Theory) 17

Chapter 2 Beam Sections Subject to Bending 19

    2.1   Introduction 19
    2.2   Basic Theory... 19
    2.3   Parallel Axis Theorem (see Figure 2.1)....................................... 21

Chapter 3 Shaft Design Basics.. 31

    3.1   Introduction 31
    3.2   Procedure for Design and Analysis of a Shaft 31
           3.2.1   Design Requirements for the Shaft 31
           3.2.2   Geometry of the Shaft.. 31
           3.2.3   Calculate the Forces Acting on the Shaft........................ 32
           3.2.4   Calculate the Bending Moments and Shear Forces Acting on
                    the Shaft 32
           3.2.5   Determine the Torsional Profile of the Shaft 34
           3.2.6   Calculate the Critical Diameters for the Shaft.................. 35

3.3   Section Modulus 36
        3.3.1   Angle of Twist 37
        3.3.2   ASME Shaft Equations 38
        3.3.3   Fillet Radii and Stress Concentrations 39
        3.3.4   Undercuts 40

Chapter 4   Combined Torsion and Bending 45

Chapter 5   Keys and Spline Calculations 57

5.1   Introduction 57
        5.1.1   Feather Key 57
        5.1.2   Straight Spline 57
        5.1.3   Involute Spline 57
5.2   Procedure for Estimating the Strength Capacity of Shaft 57
5.3   Strength Capacity of Key 58
5.4   Strength Capacity of an ISO Straight Sided Spline 60
5.5   Strength Capacity of ISO Involute Spline 60
5.6   Example Calculations 61
        5.6.1   Shaft Calculations 63
        5.6.2   Key Calculations 63
        5.6.3   Straight Spline Calculations 64
        5.6.4   Involute Spline Calculations 65

Chapter 6   Methods of Attachments 69

6.1   Bolts in Tension 69
        6.1.1   Loading Producing a Tensile Load in Bolt 69
                6.1.1.1   Permissible Stress 71
        6.1.2   Load Producing a Tension and Shear Load in Bolt 71
        6.1.3   Bolts in Shear due to Eccentric Loading 73
6.2   Welding (Permanent) 75
        6.2.1   Strength of Welded Joints 75

Chapter 7   Columns and Struts 79

7.1   Background 79
7.2   Rankine-Gordon Method 80
7.3   Perry-Robertson Method 84

Chapter 8   Eccentric Loading 87

Chapter 9   Flat Plates 91

Chapter 10   Thick Cylinders 99

Chapter 11   Energy Formulae 105

11.1   Flywheels Basics 105

Chapter 12 Gearing...115

    12.1 Spur Gearing 115
        12.1.1 Notation...115
        12.1.2 Working Stress σ_w...115
        12.1.3 Width of Teeth...115
    12.2 Bevel Gearing..122
        12.2.1 Modified Lewis Formula for Bevel Gears.......................................123

Chapter 13 Introduction to Material Selection 131

    13.1 Introduction 131
    13.2 Things to Consider 131
        13.2.1 Environment...131
        13.2.2 Strength 132
        13.2.3 Durability 132
        13.2.4 Stiffness...132
        13.2.5 Weight 133
        13.2.6 Manufacturing...133
        13.2.7 Cost..133
        13.2.8 Maintainability..133
    13.3 A Model for Material Selection..133
        13.3.1 Geometry...134
        13.3.2 Analysis...134
        13.3.3 Measurement Evaluation 134
        13.3.4 Material Selection 134
        13.3.5 Manufacturability..135
        13.3.6 Adequacy of Design 135
    13.4 A Material Database..135
        13.4.1 Paper Based Database 135
        13.4.2 Computer Based Database 135
        13.4.3 Material Classification and Coding......................................137
    13.5 Future Developments...142
        13.5.1 Knowledge Based Engineering (KBE) 142

Chapter 14 General Tables...149

Bibliography 155

Index...157

Preface

Within the UK, the Engineering Council is the regulatory body for the engineering profession to which all engineering institutions are regulated and hold the register of all practicing engineers. There are three grades of membership: Engineering Technician (Eng.Tech.), Incorporated Engineer (I.Eng.) and Chartered Engineer (C.Eng.). The Incorporated Engineer requires an education to the equivalent of a degree; the Chartered requires a minimum of a master's degree.

In recent years many institutions, including the Institution of Mechanical Engineers, have seen a considerable increase in applications for Eng.Tech. registration. These applicants may be following a work-based learning program such as an apprenticeship and are enrolled in the institution of their choice as a student member. Individuals who do not have any formal qualifications may also apply for registration by demonstrating at an interview that they have the required experience and competence through substantial working experience and by showing that they have sufficient working knowledge and understanding of the technical issues relating to their area of work.

This book has been written with these young engineers in mind, who are contemplating taking this important step and moving towards registration. The subject matter is not confined to these student engineers; it is hoped that more senior practicing engineers who are not contemplating registration will also find the subject matter useful in their everyday work as a ready reference guide.

The contents have been selected on subjects that young engineers may be expected to cover in their professional careers, and the text gives solutions to typical problems that may arise in mechanical design.

Computers are now universally used in design offices, and designers often use software without really understanding its structure or limitations. They may accept the "answer" without question and not carry out any qualification testing to verify its accuracy. The importance of carrying out these checks is stressed to ensure that mistakes are minimised.

The design examples selected are mainly static problems, and the writer has tried to give as wide a selection as possible in the space available. It was deliberated whether to include a selection of fatigue related problems, and after careful reflection the subject was considered to be beyond the scope of this book.

The subjects covered include the following:

- Introduction to stress calculations
- Beam sections subject to bending
- Shaft design basics
- Keys and spline strength calculations
- Columns and struts
- Gearing
- Introduction to material selection
- Conversions and general tables

Chapter 13, Introduction to Material Selection, has been added so that young engineers will give some thought to the materials used in terms of physical and mechanical properties. It is recommended that a personal database be built up listing these properties; this has been found by the writer to be a great asset in his own career when searching for information on this subject.

The solutions used in this book have been checked using MathCAD, and every effort has been made to ensure that the units are also coherent.

Any errors that are found will be totally my responsibility, and therefore I apologise beforehand for any made. Where errors are found, the writer will be very grateful if you, the reader, can advise me of them so that future reprints will be corrected.

I have to thank Professor Richard Dippery for his helpful comments when reading the draft copy, and I take this opportunity to also thank my wife, Eileen, for all the help and support given while writing the manuscript and to whom this book is dedicated.

Keith L. Richards

About the Author

Keith Richards is a retired Chartered Mechanical Design Engineer who has worked in the design industry for over 55 years. Initially he served an engineering apprenticeship with B.S.A. Tools Ltd., which manufactured a wide range of machine tools, including the Acme Gridley, a multi-spindle automatic lathe built under licence, and the B.S.A. single spindle automatic lathe. These were used in Britain and widely exported around the world.

On leaving the B.S.A., for a number of years he served as a freelance Engineering Designer covering a wide range of industries, including aluminium rolling mill design for installation in a company in Yugoslavia, an industrial forklift truck for an American company that was manufactured in America and Europe, and the prototype Hutton tension leg platform, an offshore oil production platform using drill string technology to anchor it to the seabed. His responsibility on this project covered the design and engineering of the mooring system components of the platform and was answerable to the customer (Conoco) and Lloyds Inspectorate for all the engineering aspects to enable the platform operators to receive the licence to operate in the North Sea.

Other work covered experimental and analytical stress analysis, photo-elastic stress analysis, residual stress determinations, and electric strain gauge analysis. One aspect of this work involved the environmental testing of specialised camera support equipment for the European Space Agency (ESA) space probe Giotto. This work was contracted to British Aerospace, which designed the support. One major problem of working in space is the very high voltages developed, and concern had been expressed that if there was an insulation breakdown in the support, then the camera would be irreparably damaged and the mission would lose the opportunity of photographing the comet's head. The probe survived and went on to investigate a further comet, Grigg-Skjellerup.

He was also involved in the design of the chassis of a vehicle to carry a 50 ton nuclear waste container, transporting it from the reactor building to the cooling ponds at Berkeley Nuclear Power Station. The design brief was that the vehicle had to be electrical/hydraulic powered and reliable, as any breakdown would create a number of problems arising from radiation due to its contents.

Other work in the nuclear industry included working with a small team at Atomic Energy Research Establishment (AERE) (Harwell) designing a hydraulic powered robotic manipulator arm, Artisan, that was used for clearing away waste from inside the nuclear storage areas at various national and international nuclear power stations. This arm was fitted with a three dimensional camera to facilitate operation of the arm from a remote position.

Keith also designed a pipeline for conveying liquid carbon dioxide from a storage area across a roadway to a vaporiser used to cool the nuclear reactors at Hinkley Point B Nuclear Power Station. This work also included designing a bridge structure for supporting the pipeline crossing the roadway. The design brief included that the pipe bridge should withstand an impact from a truck travelling at 20 miles per hour without any damage being sustained by the pipeline. Any failure in the fluid supply would cause significant inconvenience to the site operators keeping the reactors cool.

Keith was also involved in the design and manufacture of a fully automated special purpose packaging line for handling radio-active medical isotopes; these were shipped to all parts of the world. The line was designed such that the isotopes were loaded at the start of the line and finished radiation proof packages were discharged at the end of the line complete with all the necessary attachments, etc., without any human intervention. Due to the high radiation levels, human operators were only allowed into the facility for a maximum of 2 hours; hence reliability had to be a high priority.

In recent years Keith became more involved in the aerospace industry, working on projects covering aircraft undercarriages, environmental control systems for military and commercial aircraft, and the A380 wing box and trailing edge panels.

1 Introduction to Stress and Strain

This chapter is written for student engineers with only a rudimentary understanding of stresses and strains and their application to design.

The reader will be introduced to the concepts of direct stress and strain. This includes tensile, compressive and shear strains, and also defines the modulus of elasticity and rigidity.

1.1 DIRECT STRESS

When a component has either a tensile or compressive force applied to it, the component will either stretch or be squashed, and the material is then said to be stressed. Stresses cannot be measured directly; they have to be deduced from strain measurements.

The following brief notes will give some explanation to the terms used in stress calculations.

1.2 TENSILE STRESS

Consider a circular solid bar having a cross-sectional area A subject to an applied tensile force F, as shown in Figure 1.1. This force is trying to extend the bar by the dimension δ.

$$\text{Stress } \sigma = \frac{F}{A} \text{ the symbol for stress is denoted by } \sigma. \tag{1.1}$$

$$\text{Strain } \varepsilon = \frac{\delta}{L} \text{ the symbol for strain is denoted by } \varepsilon. \tag{1.2}$$

$$\text{Stiffness } K = \frac{F}{\delta} \text{ the symbol for strain is denoted by } K. \tag{1.3}$$

1.3 COMPRESSIVE STRESS

Consider the same shaft as shown in Figure 1.1, but this time the force F is now compressing the bar as shown in Figure 1.2 and shortening the bar by the dimension δ.

The fundamental unit of stress in SI units is the Pascal. In the engineering field the Pascal ($1/m^2$) is generally considered a small quantity, and therefore multiples of kPa, MPa and GPa are used.

Areas may be calculated in mm^2, and here the units of stress measured in N/mm^2 are quite acceptable. As 1 N/mm^2 is equivalent to 1,000,000 N/m^2, then it will follow that 1 N/mm^2 is the same as 1 MPa.

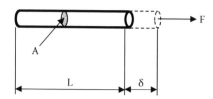

FIGURE 1.1   A circular solid bar under direct tension.

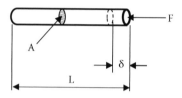

FIGURE 1.2   A circular solid bar under direct compression.

1.4   DIRECT STRAINS

In the above discussion on stress it was shown that the force F produces a deformation δ in the length of the component.

This change in length is referred to as *strain* and is defined as:

$$= \frac{300}{100}$$ The symbol for strain is ε (epsilon).

Strain has no units, as it is the ratio of the change in length to the original length, and the units therefore cancel out. Most engineering material has low strain values, as excessive strain will lead to extensive damage in the material. It will be found when studying the subject further that strain is generally written in the exponent of 10^{-6}, and this is usually written as με (micro-strain).

Example 1.1

Consider a metal rod 12.0 mm diameter and 2000 mm long subject to a tensile force of 250 N. The bar stretches 0.3 mm. Assuming the material is elastic, determine the following:

1. The stress in the rod.
2. The strain in the rod.

Solution:

Area of rod:

$$A = \frac{\pi d^2}{4}$$

$$= \frac{\pi \times 12.0^2}{4} \tag{1.4}$$

$$Area = 113.097 \text{ mm}^2$$

1. The stress in the rod:

$$\sigma = \frac{F}{A}$$

$$= \frac{250.0}{113.097} \tag{1.5}$$

$$\sigma = 2.210\,\text{N/mm}^2 \text{(or 2.21\,MPa)}$$

2. The strain in the rod:

$$\varepsilon = \frac{\delta}{L}$$

$$= \frac{0.30}{2000} \tag{1.6}$$

$$= 0.00015\,(150\,\mu\varepsilon)$$

1.5 MODULUS OF ELASTICITY (E)

When an elastic material is stretched, it will always return back to its original shape when released. Figure 1.3 shows that the deformation of the material is directly proportional to the force causing the extension. This is known as Hooke's law.

$$\text{Stiffness} = \frac{F}{\delta}$$

$$= k\,\frac{N}{m} \tag{1.7}$$

Different classes of materials will have different stiffnesses dependent upon the material and size. The size characteristic can be eliminated by using stress and strain values instead of force and deformation.

Force and deformation can be related to direct stress and strain:

$$F = \sigma \cdot A \tag{1.8}$$

$$\delta = \varepsilon \cdot L$$

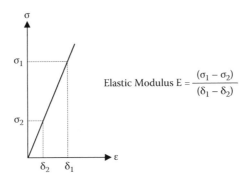

$$\text{Elastic Modulus } E = \frac{(\sigma_1 - \sigma_2)}{(\delta_1 - \delta_2)}$$

FIGURE 1.3 Relationship between stress and strain.

Therefore

$$\frac{F}{A} = \frac{\sigma A}{\varepsilon L} \qquad (1.9)$$

and

$$\frac{F \cdot L}{A \cdot \delta} = \frac{\sigma}{\varepsilon} \qquad (1.10)$$

The stiffness is in terms of stress and strain only, and this will be a constant. This constant is known as the *modulus of elasticity* and has the symbol E.

Hence:

$$E = \frac{F \cdot L}{A \cdot \delta} = \frac{\sigma}{\varepsilon} \qquad (1.11)$$

Plotting stress against strain will give a straight line having a gradient of E (see Figure 1.3). The units of E are the same as stress.

1.6 ULTIMATE TENSILE STRESS

All materials, when stretched, will reach a point when the material has deemed to have failed. This failure may be when there is a catastrophic break. This stress level is known as the *ultimate tensile stress* (UTS). Different materials will have failure values dependent upon the material type.

Example 1.2

A tensile test carried out on a steel test specimen having a cross-sectional area of 150 mm² and a gauge length of 50 mm results in the elastic section having a gradient of 500×10^3 N/mm.
    Determine the modulus of elasticity.

Solution:

From the ratio $\frac{F}{A}$ the gradient may be established, and this can be used to calculate E.

$$E = \frac{\sigma}{\varepsilon} = \frac{F}{\delta} \times \frac{L}{A}$$

$$= 500 \times 10^3 \times \frac{50}{100}$$

$$= 166.667 \text{ N/mm}^2 (166.667 \text{ MPa}).$$

1.7 SHEAR STRESS

When a force is applied transverse to the length of the component (i.e. sideways) the force is known as a shear force. Examples of this occur when a material is punched as in Figure 1.4, when a beam carries a transverse load as in Figure 1.5, or a pin is carrying a load as in Figure 1.6.

Shear stress is the force per unit area that is subject to the force as the cross-sectional area of the beam or the cross-sectional area of the pin. The unit for shear stress is τ (tau).

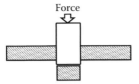

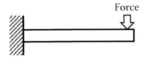

FIGURE 1.4 Material being punched.

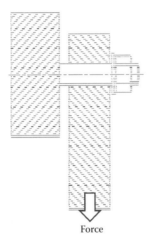

FIGURE 1.5 Beam subject to a transverse force.

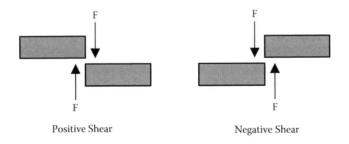

FIGURE 1.6 Pin subject to shear force.

FIGURE 1.7 Direction of shear.

$$\text{Shear stress } \tau = \frac{F}{A} \tag{1.12}$$

The sign convention for shear force and shear stress is dependent upon how the material is being sheared. Figure 1.7 defines both positive shear and negative shear.

To understand the basic theory of the shear process, consider a block of rubber that is subject to a sideways force as shown in Figure 1.8

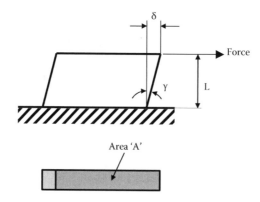

FIGURE 1.8 Block of rubber subject to sideways force.

where

  F = sideways force
  L = depth of section
  δ = shear deflection

1.8 SHEAR STRAIN

As in Figure 1.8 the force F causes the block to deform. The shear strain is defined as the ratio of the height L to the distance deformed δ, i.e. δ/L.

It is also seen in Figure 1.8 that the end face rotates through an angle γ; as this is generally a very small angle, it can be considered that the distance δ is the length of an arc having a radius of L with an angle γ such that:

$$\gamma = \frac{\delta}{L} \qquad (1.13)$$

The symbol for the shear strain is γ (gamma).

1.9 MODULUS OF RIGIDITY

Just as the modulus of elasticity, E, relates tensile stress to tensile strain, the modulus of rigidity, G, relates shear stress to shear strain, and a plot of this relationship will give a straight line as shown in Figure 1.9.

The gradient of the line is constant $\frac{F}{\delta}$, and this is the spring stiffness of the block of rubber in N/m. Other materials will display different spring stiffnesses.

If the force F is divided by the area A and δ by the height L, the relationship will still be a constant such that:

$$\frac{F}{A} \div \frac{\delta}{L} = \frac{F \cdot L}{A \cdot \delta} = \text{constant} \qquad (1.14)$$

Now:

$$\frac{F}{A} = \tau \text{ and } \frac{\delta}{L} = \gamma \qquad (1.15)$$

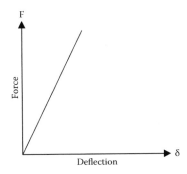

FIGURE 1.9 Modulus of rigidity.

Hence:

$$\frac{F \cdot L}{A \cdot \delta} = \frac{\tau}{\gamma} = \text{constant} \qquad (1.16)$$

This constant is known as the *modulus of rigidity* and has the symbol G.

1.10 ULTIMATE SHEAR STRESS

Permanent deformation will occur in a material if the material is sheared beyond a certain limit and does not return back to its original shape. In this instance the elastic limit has been exceeded. When the material is stressed to the limit where the part fractures into two separate pieces, i.e. in a punching operation or a pin joint fails, the *ultimate shear stress* has been reached. The ultimate shear stress has the symbol τ_u.

Example 1.3

Calculate the force required to pierce a hole 20.0 mm diameter in a sheet 5.0 mm thick given that the ultimate shear stress is 50.0 MPa.

Solution:

The area to be pierced:
   Circumference of cut:

$$\pi \cdot D = \pi \times 20.0 \text{ mm}$$

$$= 62.832 \text{ mm}$$

Area of cut:

$$= 62.832 \times 5.0 \text{ mm}^2$$

$$= 314.159 \text{ mm}^2$$

The ultimate shear strength = 50 N/mm²:

$$\tau = \frac{F}{A} \therefore F = t \cdot A$$

Shear force required:

$$F = 1256.64 \text{ kN}$$

1.11   DOUBLE SHEAR

Consider a pinned joint as shown in Figure 1.10 that is supported at each end of the pin. This type of joint is known as a *Clevis and Clevis pin*. In this type of joint the shear area will be double that of a single pin or rivet in a thickness of material.

From Figure 1.10 it is seen that by the balance of forces, the force in each of the supports will be F/2.

As the shear area is twice the cross-sectional area of the pin diameter, it will require double the amount of force to shear the pin as it would have done for a pin in single shear.

Example 1.4

A pin is used to attach a clevis on a cylinder to an adjacent fitting. The cylinder force is 120 kN and the maximum shear stress allowed for the clevis pin is 75 MPa.

Determine the diameter of a suitable pin size.

Solution:

As the pin is in double shear, the shear stress:

$$\tau = \frac{F}{2A}$$

The shear area of the pin:

$$A = \frac{F}{2t} = \frac{120 \times 10^3}{2 \times 75 \times 10^6}$$

$$A = 800 \times 10^6 \text{ m}^2$$

$$A = 800 \text{ mm}^2 = \frac{\pi \cdot d^2}{4}$$

$$d = \sqrt{\frac{4 \times 800}{\pi}} = 31.92 \text{ mm}$$

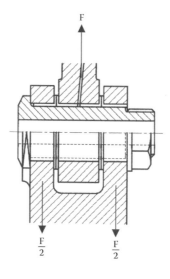

FIGURE 1.10   Pinned lug.

1.12 POISSON'S RATIO

When a strip of isotropic material is stretched in one direction (say the y direction), there will be a corresponding contraction in another direction (x direction). This is illustrated in Figure 1.11. The positive strain in the y direction will be ε_y, and the strain in the other two directions will be $-\varepsilon_x$, i.e. the strain will be negative.

For elastic materials it is found that the applied strain (ε_y) is always directly proportional to the induced strains ε_x such that:

$$\frac{\varepsilon_x}{\varepsilon_y} = -\nu \tag{1.17}$$

ν (nu) is an elastic constant called Poisson's ratio.

Hence the strain produced in the x direction will be $\varepsilon_x = -\nu\varepsilon_y$.

If the stress is applied in the x direction, the resulting strain in the y direction will be:

$$\varepsilon_y = -\nu\varepsilon_x \tag{1.18}$$

Conversely, if the strip is compressed in the y direction, there will be a corresponding expansion in the other directions.

Now consider if the material has an applied stress in both the y and x directions as shown in Figure 1.12.

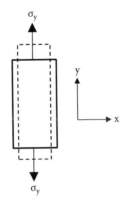

FIGURE 1.11   Poisson's ratio, y direction.

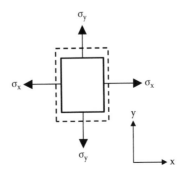

FIGURE 1.12   Poisson's ratio, x and y directions.

The resulting strain in any one direction will be the sum of the strains due to the direct force and the induced strain from the other direct force.

Hence:

$$\varepsilon_x = \frac{\sigma_x}{E} - \nu\sigma_y = \frac{\sigma_x}{E} - \nu\frac{\sigma_y}{E}$$

$$\varepsilon_x = \frac{1}{E}(\sigma_x - \nu\sigma_y)$$

(1.19)

Similarly:

$$\sigma = \frac{F}{A}\varepsilon_y = \frac{1}{E}(\sigma_x - \nu\sigma_y)$$

(1.20)

Example 1.5

A material element has stresses of 3.0 MPa applied in the x direction and 5.0 MPa in the y direction. The elastic constants for the material are E = 205 GPa and ν = 0.27.

Determine the strains in both the x and y directions.

Solution:

$$\varepsilon_x = \frac{1}{E}(\sigma_x - \nu\sigma_y)$$

$$= \frac{1}{205 \times 10^9}(3.0 \times 10^6 - 0.27 \times 5.0 \times 10^6)$$

$$= 8.049 \ \mu\varepsilon$$

$$\varepsilon_y = \frac{1}{E}(\sigma_y - \sigma_x)$$

$$= \frac{1}{205 \times 10^9}(5.0 \times 10^6 - 0.27 \times 3.0 \times 10^6)$$

$$= 20.440 \ \mu\varepsilon$$

Note: The above solutions are not restricted to the x and y directions. The formula works for any two orthogonal stresses. σ_1 and σ_2 with the corresponding strains ε_1 and ε_2 are generally used.

1.13   CONVERTING BETWEEN STRESSES AND STRAINS

Having derived:

$$\varepsilon_1 = \frac{1}{E}(\sigma_1 - \nu\sigma_2) \text{ and } \varepsilon_2 = \frac{1}{E}(\sigma_2 - \nu\sigma_1)$$

(1.21)

and combining and rearranging to make σ_2 the subject:

$$\sigma_2 = \varepsilon_2 E + \nu\sigma_1$$

(1.22)

Substituting this result into the first equation:

$$\varepsilon_2 = \frac{(\varepsilon_2 E + \nu\sigma_1) - \nu\sigma_1}{E}$$

(1.23)

and rearranging to make σ_1 the subject:

$$\sigma_1 = \left(\frac{E}{1-v^2}\right)(\varepsilon_1 + v\varepsilon_2) \qquad (1.24)$$

Repeat and make σ_2 the subject:

$$\sigma_2 = \left(\frac{E}{1-v^2}\right)(\varepsilon_2 + v\varepsilon_1) \qquad (1.25)$$

1.14 THREE DIMENSIONAL STRESS AND STRAIN

The above equations (1.24 and 1.25) were derived for a two dimensional system. It is now possible to extend the solution to cover a material that is stressed in three mutually perpendicular directions x, y and z. The strains in any one of these axes will be reduced by the effect of the strain in any of the other two directions.

The three strains are:

$$\varepsilon_1 = \frac{1}{E}(\sigma_1 - v\sigma_2 - v\sigma_3) = \frac{1}{E}[\sigma_1 - v(\sigma_2 + \sigma_3)] \qquad (1.26)$$

$$\varepsilon_2 = \frac{1}{E}(\sigma_2 - v\sigma_1 - v\sigma_3) = \frac{1}{E}[\sigma_2 - v(\sigma_1 + \sigma_3)] \qquad (1.27)$$

$$\varepsilon_3 = \frac{1}{E}(\sigma_3 - v\sigma_1 - v\sigma_2) = \frac{1}{E}[\sigma_3 - v(\sigma_1 + \sigma_2)] \qquad (1.28)$$

1.15 VOLUMETRIC STRAIN

If a cube of material that is stressed in the x direction by a compressive pressure as shown in Figure 1.13 is considered, the change in volume will be:

$$L^2\Delta L \qquad (1.29)$$

If the cube is now considered to be hydrostatically strained by an equal amount in the y and z directions as well, with very little error the total change in volume will be:

$$3L^2\Delta L \qquad (1.30)$$

where the original volume is L^3.

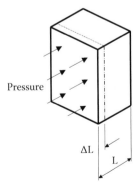

FIGURE 1.13 Volumetric strain.

When a solid object is subjected to a pressure p causing the volume to be reduced, the volumetric strain will be:

$$\varepsilon_v = \frac{\text{change in volume}}{\text{original volume}} \qquad (1.31)$$

In the case of a cube:

$$\varepsilon_v = \frac{3L^2\Delta L}{L^3} \qquad (1.32)$$

$$= \frac{3\Delta L}{L}$$

$$\varepsilon_v = 3\varepsilon \qquad (1.33)$$

ε is the equal strain in all three directions.

It follows that when a material is compressed by a pressure which by definition must be equal in all three directions, the volumetric strain is three times the linear strain in any direction.

1.16   BULK MODULUS

The volumetric strain in an elastic material is directly proportional to the stress producing that strain such that:

$$K = \frac{\sigma}{\varepsilon_v} \qquad (1.34)$$

K is the symbol for the bulk modulus and is considered a further material constant.

1.17   RELATIONSHIP BETWEEN THE ELASTIC CONSTANTS

When a material is compressed by a pressure p, the stress is obviously equal to −p as it is compressive.

The bulk modulus becomes:

$$K = \frac{-p}{\varepsilon_v} \qquad (1.35)$$

From Equation (1.33)

$$\varepsilon_v = 3\varepsilon$$

Then from Equations (1.27), (1.28) and (1.29), the strains in all three directions being equal and the stresses being equal to −p:

$$\varepsilon = \frac{1}{E}(-p + v2p) \qquad (1.36)$$

The volumetric strain then becomes:

$$\varepsilon_v = 3\varepsilon = \frac{3}{E}(-p + v2p) \qquad (1.37)$$

Now combining Equations (1.36) and (1.37):

$$K = \frac{-p}{\left(\dfrac{3}{E}\right)(-p + v2p)} \qquad (1.38)$$

$$= \frac{E}{3(1-2v)} \tag{1.39}$$

This shows the relationship between K, E and v.

The relationship between the shear modulus G and the other elastic constants is given by:

$$G = \frac{E}{2(1+v)} \tag{1.40}$$

This equation is given here without any proof. Further study into three dimensional systems will be required but is considered too advanced for this particular work.

1.18 FACTOR OF SAFETY IN TENSILE OR COMPRESSIVE MODE

When designing a component or structure that will be subject to loading, a Factor of Safety (FoS) has to be considered to ensure that the working stresses are kept within safe limits.

For brittle materials the FoS is defined in terms of the ultimate tensile strength:

$$\text{FoS} = \frac{\text{Ultimate tensile stress}}{\text{Maximum working stress}} \tag{1.41}$$

For ductile materials the FoS is more usually defined in terms of the yield stress. The yield stress is the value of stress when the material changes from elastic to plastic.

$$\text{FoS} = \frac{\text{Yield stress}}{\text{Maximum working stress}} \tag{1.42}$$

Determining the FoS: The FoS chosen will be dependent upon a range of conditions relating to the function of the component or structure when in service. Some of the conditions are listed below:

- Potential overloads
- Possible defects in the materials used
- Defects in workmanship
- Deterioration due to wear, corrosion, etc.
- The possibility of any load being applied suddenly or repeated loading

Example 1.6

A tie bar is manufactured from a material having an ultimate tensile strength of 600 MPa. Determine the maximum safe working stress if a FoS of 4 is used.

Solution:

$$\text{Maximum working stress} = \frac{\text{Ultimate tensile stress}}{\text{Factor of Safety}}$$

$$\varepsilon_x = \frac{\sigma_x}{E} - v\sigma_y = \frac{\sigma_x}{E} - v\frac{\sigma_y}{E} = \frac{600 \text{ MPa}}{4}$$

$$= 150 \text{ MPa}$$

1.19   FACTOR OF SAFETY IN SHEAR MODE

FoS can also be used when in shear mode; here, instead of the UTS being used, the material's ultimate shear stress is used:

$$\text{Factor of Safety} = \frac{\text{Ultimate shear stress}}{\text{Maximum working shear stress}} \qquad (1.43)$$

Example 1.7

A shear pin is manufactured from a material with an ultimate shear stress of 300 MPa and a maximum working shear stress not to exceed 100 MPa; calculate the FoS.

Solution:

$$\text{Factor of Safety} = \frac{\text{Ultimate shear stress}}{\text{Maximum working shear stress}}$$

$$= \frac{300}{100}$$

$$= 3.0$$

1.20   THEORIES OF ELASTIC FAILURE

To complete this chapter a brief discussion on theories of failure will not be out of place. Although not considered in the remaining chapters, it is felt that the student should be aware of them for future study.

In a simple tensile test elastic failure is assumed to occur when the stress in the specimen reaches the elastic limit stress for the material. This stress will be denoted by σ_0.

There are a number of theories, but this section will confine its attention to the following three theories.

- Maximum principal stress theory (Rankine)
- Maximum principal strain theory (St. Venant)
- Shear strain energy theory (Von Mises)

At this point what is regarded as a failure? Failure could be regarded as when the material fractures or when the material permanently yields.

If a simple tensile test on a material is carried out, the resultant stress-strain curve may look like the one shown in Figure 1.14. The maximum allowable stress in the material is σ_{max}. This may be regarded as the stress at fracture (UTS), the stress at the yield point or the stress at the limit of proportionality (often the same as the yield point if the material displays a definite yield point such as steel). The modulus of elasticity has been defined as $E = \frac{\sigma}{\varepsilon}$, and this is only true up to the limit of proportionality.

There is only one direct stress in the tensile test $\sigma = \frac{F}{A}$, so that it follows that $\sigma_{max} = \sigma_1$ and the corresponding strain $\varepsilon_{max} = \varepsilon_1$. Complex stress theory states that the maximum stress (τ) and strain (γ) will occur on a plane at 45° to the principal plane.

When studying the fracture point on a failed ductile specimen, a cup and cone are formed with sides at 45° to the centre axis with a small amount of necking. Brittle materials often fail without displaying any necking but with a flat fail plane at 45° to the axis. This will suggest that fracture occurred due to the maximum shear stress being reached. Figure 1.15 illustrates a tensile test failure of a ductile specimen.

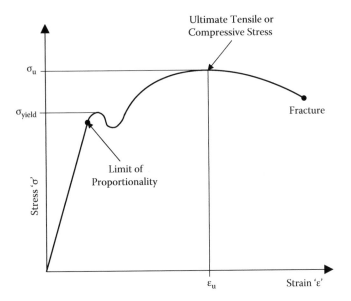

FIGURE 1.14   Typical stress-strain curve.

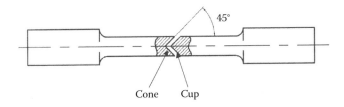

FIGURE 1.15   Tensile test failure.

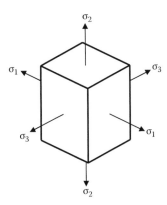

FIGURE 1.16   Volumetric strain.

In a complex stress situation there are three principal stresses acting on the body, σ_1, σ_2 and σ_3. σ_1 is the greatest and σ_3 will be the smallest. There will be corresponding principal strains ε_1, ε_2 and ε_3, together with shear strains.

Figure 1.16 shows a three dimensional cube with the three principal stresses applied.

To minimise the complexity of the calculations, only the two dimensional cases will be shown, i.e. $\sigma_3 = 0$.

1.20.1 Rankine's Principal Stress Theory

This theory simply states that in a complex stress situation the material fails when the greatest principal stress equals the maximum allowable value:

$$\sigma_0 = \sigma_{max}$$

σ_{max} could be the stress either at yield or fracture depending upon the definition of failure. If σ_1 is less than σ_{max}, then the material can be considered safe.

The safety factor can be expressed as:

$$\text{Factor of Safety (FoS)} = \frac{\sigma_{max}}{\sigma_1} \qquad (1.44)$$

Example 1.8

A specimen fractures at a stress level of 950 MPa during a tensile test. A safety factor of 5 is required when the material is used as part of a structure. Determine the greatest principal stress that should be allowed to occur when failure is based on the Rankine theory.

Solution:

$$\text{FoS} = 5 = \frac{\sigma_{max}}{\sigma_1}$$

$$= \frac{950 \text{ Mpa}}{\sigma_1}$$

Hence:

$$= \frac{950 \text{ MPa}}{5}$$

$$= 190 \text{ MPa}$$

1.20.2 St. Venant's Maximum Principal Strain Theory

St. Venant's theory predicts that failure will occur when the greatest principal strain reaches the strain at the elastic limit in a simple tensile test, i.e.:

$$\sigma_0 = \sigma_x - \nu\sigma_y \qquad (1.45)$$

For like stresses, the theory will give $\sigma_x > \sigma_0$, but this is not substantiated by experiment and this theory finds little general support.

Example 1.9

A candidate material fails in a simple tensile test at a stress level of 600 MPa. The designated material is to be used in a loaded structure and must have principal stresses of 600 and 400 MPa. Establish the FoS at this load based on the maximum principal strain theory. Take Poisson's ratio as 0.28.

Solution:

$$\text{FoS} = \frac{\sigma_{max}}{(\sigma_1 - \nu\sigma_2)}$$

$$= \frac{600}{600 - 0.28 \times 400}$$

$$= 1.230$$

The component will be safe as the factor of safety is greater than 1.

1.20.3 SHEAR STRAIN ENERGY THEORY (VON MISES THEORY)

Von Mises theory is the most common failure theory used in engineering and possibly the most accurate. It is generally used in three dimensional stress analyses for complex stress loading conditions. Within the component there are stresses acting in different directions, and the direction and magnitude of these stresses change from point to point. The Von Mises criterion is a formula for calculating whether these stress combinations at any given point will result in a failure.

The theory is used for ductile materials and can be utilised for evaluating stresses both static and dynamic. It is able to combine principal stresses from the Mohr's circle covering bending and torsion into an equivalent applied stress which can then be compared to the allowable stress of the material.

Although most engineers refer to it as Von Mises stress, the correct title is *Von Mises–Hencky criterion for ductile failure*.

Von Mises found that even though none of the principal stresses exceeded the yield stress of the material, it is possible for yielding to occur from the combination of these stresses.

The equivalent stress is often referred to as the *Von Mises stress* as a shorthand description, although it is not strictly a stress. It is a number that can then be used as an index. If the index exceeds the yield stress, then the material is considered to be in a failure condition.

It is also sometimes referred to as the *distortion-energy theory*.

For a two dimensional problem the following equation is applicable:

$$\sigma_o^2 = \sigma_x^2 + \sigma_y^2 - \sigma_x \sigma_y \qquad (1.46)$$

Example 1.10

Consider the question set out in Example 1.9.

Solution:

$$\sigma_o^2 = \sigma_x^2 + \sigma_y^2 - \sigma_x \sigma_y$$

$$= 600^2 + 400^2 - (600 \times 400)$$

$$= 280 \times 10^3$$

Therefore $\sigma_o = 529.15$ MPa.

The FoS will then be:

$$FoS = \frac{\sigma_{max}}{\sigma_o}$$

$$= \frac{600 \text{ MPa}}{529.15 \text{ MPa}}$$

$$FoS = 1.134$$

From the above it is seen that although the FoS is still greater than 1.0 the Von Mises theory is more conservative than using the Rankine theory.

2 Beam Sections Subject to Bending

2.1 INTRODUCTION

For the purposes of this chapter, a beam is considered to be a structural member having a constant cross section and loaded in the transverse direction to its length.

It does not include being subject to any axial or torsion loads.

Beams can be classified into basically four groups (see Table 2.1):

1. Cantilever (built in at one end only).
2. Propped cantilever (built in at one end and simply supported at the free end).
3. Horizontal beam simply supported at each end.
4. Horizontal beam in encastré (built in at both ends).

Table 2.2 shows some standard loading cases for horizontal beams subject to various loadings.

Table 2.3 shows some common sectional properties.

Beams are generally considered to be horizontal or near horizontal. In the case of the beam being vertical and subject to axial loads, these are known as columns and struts and are subject to a different form of analysis (these are considered in Chapter 7).

2.2 BASIC THEORY

From the bending theory:

$$\frac{M}{I} = \frac{\sigma}{y} = \frac{E}{R} \tag{2.1}$$

where

M = maximum bending moment (Nm)
I = second moment of area of section about neutral axis NA (m^4)
σ = extreme fibre stress (Pa)
y = distance from neutral axis to extreme fibre (m)
E = modulus of elasticity (Pa)
R = radius of curvature (m)

Also,

$$M = \sigma \cdot Z \tag{2.2}$$

where $Z = I/y$ and is known as the *section modulus*.

Various values of I for a number of sections can be found in a number of steel manufactures or steel stockists catalogues; there may also be other tables.

TABLE 2.1
Standard Bending Cases Considered in This Chapter

Case	Description	Graphic
1.	**Cantilever** With a concentrated load at the free end	
2.	**Propped cantilever** With a concentrated load at the centre of span	
3.	**Simply supported at each end** With a concentrated load at the centre of span	
4.	**Encastré** With a concentrated load at the centre of span	

TABLE 2.2
Standard Loading Cases for Horizontal Beams

Case No.	Description	Graphic	Moment 'M' Max	Deflection 'δ' Max
1.	Cantilever with concentrated load at the free end		WL	$\dfrac{WL^3}{3EI}$
2.	Cantilever with an intermediate load		Wa	$\dfrac{Wa^3}{3EI}\left(1+\dfrac{3b}{2a}\right)$
3.	Cantilever with uniformly distributed load		$\dfrac{wL^2}{2}$	$\dfrac{wL^4}{8EI}$
4.	Propped cantilever with a central load		$Ma=\dfrac{Wb(L^2-b^2)}{2.L^2}$ $Mc=\dfrac{Wb}{2}2-\left(2-\dfrac{3b}{L}+\dfrac{b^3}{L^3}\right)$	$\dfrac{Wa^3b^2}{12EIL^3}(4L-a)$
5.	Popped cantilever with an intermediate load		$Ma=\dfrac{3W^2}{16}$ $Mc=\dfrac{5WL}{32}$	$0.000932\dfrac{WL3}{3EI}$
6.	Propped cantilever with a uniformly distributed load		$\dfrac{WL}{8}$	$\dfrac{WL^3}{185EI}$
7.	Beam supported at two points with single load at centre of span		$\dfrac{WL}{4}$	$\dfrac{WL^3}{48EI}$

TABLE 2.2 (*Continued*)
Standard Loading Cases for Horizontal Beams

Case No.	Description	Graphic	Moment 'M' Max	Deflection 'δ' Max
8.	Beam supported at two points with an intermediate load		$\dfrac{Wab}{L}$	$\dfrac{WL^3}{48EI}$ $\left[\dfrac{3a}{L}-4\left(\dfrac{a}{L}\right)^3\right]$
9.	Beam supported at two points with uniformly distributed load along its length		$\dfrac{wL^2}{8}$	$\dfrac{5wL^4}{384EI}$
10.	Beam in encastré with a central load		$\dfrac{WL}{8}$	$\dfrac{WL^3}{192EI}$
11.	Beam in encastré with an intermediate load		$M_A=-\dfrac{Wab^2}{L^2}$ $M_B=-\dfrac{Wba^2}{L^2}$ $M_C=\dfrac{2Wa^2b^2}{L^3}$	$\dfrac{2Wa^2b^2}{3EI(3L-2a)^2}$
12.	Beam in encastré with a uniformly distributed load		$\dfrac{wL^2}{12}$	$\dfrac{wL^4}{384EI}$

Where: W = total load (N)
     w = uniformly distributed load (N per m)
     L = effective span (m)

In a number of cases, the value for I will require calculation; this may necessitate that the section be split up into various calculable parts using the *parallel axis theorem*.

2.3 PARALLEL AXIS THEOREM (SEE FIGURE 2.1)

The second moment of area of a section about an axis x:x parallel to an axis n:a passing through the centroid of the section is given by:

$$I_{xx} = I_{na} + Ah^2 \tag{2.3}$$

where
     I_{xx} = second moment about XX axis (m⁴)
     I_{na} = second moment of section about the centre of gravity (CG) (m⁴)
     A = area of section (m²)
     h = the perpendicular distance between x:x axis and CG (m)

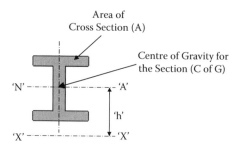

FIGURE 2.1   Parallel axis theorem.

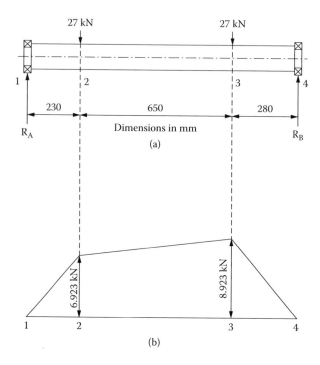

FIGURE 2.2   (a) Loaded beam. (b) Bending moment diagram.

Example 2.1

A mild steel shaft is loaded as shown in Figure 2.2(a). Construct a bending moment diagram and determine the maximum diameter of the shaft required in the central portion of the shaft if the ultimate tensile strength (σ_t) of the material is limited to 460 MPa.

Consider if a Safety Factor (SF) of 5 is appropriate.

Solution:

Taking moments about R_B and bending moment at positions 2, 3 and 4, respectively,

$$M_2 = (R_A \times 0.23 \text{ m})$$

$$M_3 = (R_A \times 0.88 \text{ m}) - (27 \text{ kN} \times 0.65 \text{ m})$$

$$M_4 = (R_A \times 1.16 \text{ m}) - (27 \text{ kN} \times 0.93 \text{ m}) - (35 \text{ kN} \times 0.28 \text{ m})$$

Substituting the value for R_A results in the following moments:

$$M_2 = (30.095 \times 10^3 \times 0.23) = 6.922 \text{ kNm}$$

$$M_3 = (30.095 \times 10^3 \times 0.88) - (27 \times 10^3 \times 0.65) = 8.934 \text{ Nm}$$

$$M_4 = (30.095 \times 10^3 \times 0.16) - (27 \times 10^3 \times 0.93) - (35 \times 10^3 \times 0.28) = 0.00 \text{ kNm (check)}$$

The completed bending moment diagram is as shown in Figure 2.2(b).
From the bending formula:

$$\frac{M}{I} = \frac{\sigma}{y} \quad \text{(from equation 2.1)}$$

Rearrange the formula to solve for σ:

$$\sigma = \frac{M \cdot y}{I}$$

where maximum bending moment M = 8.933 kN.m.

Second moment of area $\quad I = \dfrac{\pi \times d^4}{64} \quad$ (Table 2.3 case 1)

$$y = \frac{d}{2}$$

Hence:

$$\sigma_t = \frac{8.933 \text{ kN.m} \cdot x \cdot d \times 64}{\pi \times d^4}$$

$$\sigma_t = 181.982 \times 10^3 \times d^3$$

Solving for d:

$$d = \sqrt[3]{\frac{181.982 \times 10^3 \text{ N.m}}{460 \text{ MPa}}}$$

$$d = 73.41 \text{ mm}$$

Now with a SF of 5, the ultimate tensile stress is factored by the SF.

$$\sigma_{allowable} = \frac{460 \text{ MPa}}{5 \text{ (SF)}}$$

Therefore

$$\sigma_{allowable} = 92 \text{ MPa}$$

Hence, recalculating, the revised diameter will be

$$d = 125.5 \text{ mm}$$

TABLE 2.3

Some Common Sectional Properties

Case No.	Graphic	A	A_S	I_{xx}	Z_{xx}
1		$\dfrac{\pi}{4}D^2$	$\dfrac{3}{4}A$	$\dfrac{\pi}{64}D^4$	$\dfrac{\pi}{32}D^3$
2		$\dfrac{\pi}{4}\left(D^2 - d^2\right)$	$\dfrac{D^2}{6}\left(B - b\left(t - \dfrac{2T}{D}\right)^3\right)$	$\dfrac{\pi}{64}\left(D^4 - d^4\right)$	$\dfrac{\pi}{32D}\left(D^4 - d^4\right)$
3		πDT	$\dfrac{A}{2}$	$\dfrac{\pi}{8}D^3 t$	$\dfrac{D^2 . t}{8}$
4		$B.D$	$\dfrac{2}{3}A$	$\dfrac{1}{12}B.D^3$	$\dfrac{1}{6}B.D^2$
5		$B.D - b.d$	$\dfrac{2}{3}\dfrac{B.D^3 - b.d^3}{D^2}$	$\dfrac{1}{12}\left(B.D^3 - b.d^3\right)$	$\dfrac{1}{6D}\left(B.D^3 - b.d^3\right)$
6		$B.D - b.d$	$d(B - b)$	$\dfrac{1}{12}\left(B.D^3 - b.d^3\right)$	$\dfrac{1}{6D}\left(B.D^3 - b.d^3\right)$
7		$D.t + 2b.T$	$t(D - 2T)$	$\dfrac{D^3}{12}\left(B - b\left(t - \dfrac{2T}{D}\right)^3\right)$	$\dfrac{D^2}{6}\left(B - b\left(t - \dfrac{2T}{D}\right)^3\right)$

A = cross sectional area

A_s = effective cross sectional area for the calculation of shear stress across the section

I_{xx} = second moment of area about x:x

$Z_{xx} = I_{xx/y}$ modulus of section about x:x

Note: The question was asked to consider if a SF of 5 is appropriate. It is assumed that the load-ing on the shaft is steady without any sudden impacts. In this instance, if the shaft rotates, the stresses on either side of the shaft will alter from tensile to compressive, giving rise to a fluctuat-ing stress which could eventually lead to a fatigue failure of the shaft. As a rule of thumb, provid-ing that for *steel* components the maximum working stress is kept below the *fatigue limit*, the component should not suffer any long term fatigue problems; hence this value would normally be below 40% of the ultimate tensile stress. Therefore a SF of 5 in this instance is considered appropriate.

If the material of the shaft had been, say, aluminium, then for this material there is no limiting stress, and hence the number of cycles will need to be carefully considered. The Factor of Safety (FoS) will then have to be calculated based on the fatigue strength required.

Example 2.2

The cast iron beam shown in Figure 2.3 carries a distributed load of 50 N/mm over its simply supported span of 3000 mm. Calculate the maximum and minimum stresses in the top and bottom flanges together with the maximum deflection of the beam.

Solution:

The first part of the answer is to establish the position of the neutral axis by taking moments about the edge x:x.

To answer this question, the section is broken down into smaller sections as shown in Figure 2.3. The sectional properties of the individual sections are then calculated and summed as in Table 2.3. From this table the position of the neutral axis of the individual sections can then be calculated.

The first part of the answer is to break the overall section into simple shape sections as depicted in Figure 2.3. A boundary edge is selected and used as a datum for the subsequent calculations. A table can be constructed as shown in Table 2.4; this will be used to identify the position of the neutral axis of the combined section.

The construction of the table may be achieved using hand calculations, although a spreadsheet program such as Microsoft Excel® may be more convenient, particularly where there is a large number of sections. In this section, as it is symmetrical, the centre of gravity will lie on the vertical axis; therefore only the vertical position of the neutral axis will be considered. The elements of the table are considered self-explanatory.

The position of the neutral axis from the datum edge (X:X) is calculated by multiplying the distances of the CG of the individual sections (y) from the datum by the corresponding area (column 4) for that section. The result of this calculation (Ay) is entered in column 6 of Table 2.4.

Columns 4 and 6 in Table 2.4 are then summed to give values of ΣA and ΣAy.

The value of y1 is then calculated by dividing ΣAy by ΣA, and y2 is calculated by subtracting y1 from the total vertical dimension for the figure.

The calculation for the moment of inertia (I) of the individual sections can be completed.

In Table 2.5, the distance from the 'c of g' of the individual sections to the neutral axis is then determined and this is shown in column 6.

Column 4 in Table 2.5 calculates the moments of inertia for the individual sections using the formula shown in Table 2.3. These values are the individual moment of inertia that passes through the CG.

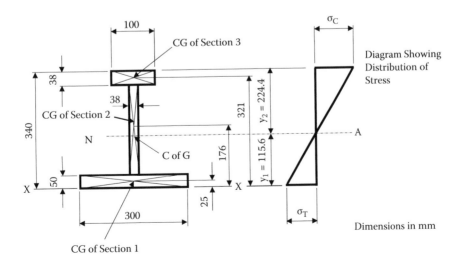

FIGURE 2.3 Example 2.2: Section beam.

TABLE 2.4
Position of Neutral Axis for Example 2.2

1	2	3	4	5	6
Section	b mm	d mm	Area mm^2	y mm	Ay mm^3
1	300	50	15000	25	0.375×10^6
2	38	252	9576	176	1.685×10^6
3	100	38	3800	321	1.220×10^6
		$\Sigma A =$	28376	$\Sigma Ax =$	3.280×10^6

$$y_1 = \frac{\Sigma Ax}{\Sigma A} = 115.60 \text{ mm}$$

$$y_2 = 340 - 115.6 = 224.4 \text{ mm}$$

TABLE 2.5
Individual and Combined Inertia Calculations for Example 2.2

1	2	3	4	5	6	7
Section	b mm	d mm	bd^3/12 mm^4	Area mm^2	h mm	$I_{xx} + (Ah^2)$ mm^4
1	300	50	3.125×10^6	15000	90.6	126.250×10^6
2	38	252	50.676×10^6	9576	60.4	85.611×10^6
3	100	38	0.4573×10^6	3800	205.4	160.78×10^6
					$\delta I_{na} =$	372.64×10^6

In column 7 of Table 2.5, the parallel axis theorem ($I_{xx} + (Ah^2)$) is now invoked and the results for the individual sections are entered. The column is summed and this gives the I_{na} for the combined section.

The maximum tensile strength in the bottom flange:

$$\sigma_{tensile} = \frac{M. y_1}{I_{xx}}$$

$$\sigma_{tensile} = \frac{56.25 \text{ MN.mm} \times 115.6 \text{ mm}}{372.64 \times 10^6 \text{ mm}^4}$$

$$\sigma_{tensile} = 17.45 \text{ MPa}$$

The maximum compressive stress in the top flange:

$$\sigma_{compressive} = \frac{M. y_2}{I_{xx}}$$

$$\sigma_{compressive} = \frac{56.25 \text{ MN.mm} \times 224.4 \text{ mm}}{372.64 \times 10^6 \text{ mm}^4}$$

$$\sigma_{compressive} = 33.873 \text{ MPa}$$

The maximum deflection in the beam will be as follows. From Table 2.2, Case 9, the maximum deflection can be calculated from the formula:

$$\delta_{max} = \frac{5wL^4}{384EI}$$

$$= \frac{5 \times 50 \text{ N/mm} \times (3000 \text{ mm})^4}{384 \times 208 \times 10^9 \text{ Pa} \times 372.64 \times 10^6 \text{ mm}^4}$$

$$\delta_{max} = 0.680 \text{ mm}$$

This deflection will occur at mid-span of the beam, i.e. 1500 mm from either end.

Example 2.3

A box girder 250 mm deep is fabricated as shown in Figure 2.4. The beam is uniformly loaded over a simple span of 6000 mm. The ultimate tensile stress of the material is 430 MPa; use a SF of 5 to find the value of the uniformly distributed load the beam will carry. Also calculate the maximum deflection in the section.

Solution:

Extract data from structural steel manufacturer's catalogue.

178 × 76 rolled steel channel:

$$\text{Area} = 26.54 \text{ cm}^2 \ (2654.0 \text{ mm}^2)$$

$$\text{Second moment of area} = 134 \text{ cm}^4 \ (1.34 \times 10^6 \text{ mm}^4)$$

70 × 70 × 10 rolled steel equal angle:

$$\text{Area} = 13.1 \text{ cm}^2 \ (1310.0 \text{ mm}^2)$$

$$\text{Second moment of area} = 57.2 \text{ cm}^4 \ (0.5720 \times 10^6 \text{ mm}^4)$$

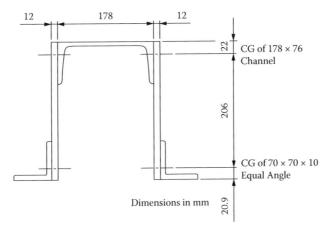

FIGURE 2.4 Example 2.3: Beam section.

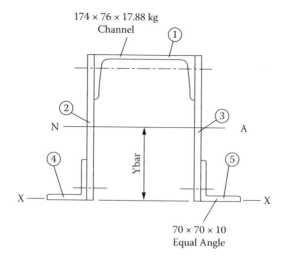

FIGURE 2.5   Identification of beam sections.

Side plates:

$$Area = 29.868 \text{ cm}^2 \ (2986.8 \text{ mm}^2)$$

$$Second \ moment \ of \ area = 1562.5 \text{ cm}^4 \ (15.625 \times 10^6 \text{ mm}^4)$$

All structural steel catalogues and steel designer manuals quote the physical properties of the sections in cm rather than mm. It will be found to be more convenient to convert to mm when constructing a table to calculate the positions of the neutral axes and second moment of areas.

First identify the individual elements as shown in Figure 2.5 so that a table can now be constructed to calculate the physical properties of the total section.

To find the maximum load (w) from the bending formula:

$$M = \frac{wL^2}{8} = \frac{I\sigma}{y} \tag{2.4}$$

Re-arranging to solve for 'w'

$$w = \frac{8 . I . \sigma}{L^2 . y . SF}$$

Inputting the following values:

$$I = 90.283 \times 10^6 \text{ mm}^4$$

$$\sigma = 430 \text{ MPa}$$

$$L = 6000 \text{ mm}$$

$$y = 125.055 \text{ mm}$$

$$SF = 5$$

gives:

$$w = 13.797 \text{ N/m}$$

TABLE 2.6
Physical Properties for Example 2.3

1	2	3	4	5	6	7	8	9
Section	b mm	d mm	A mm²	x mm	Ax mm³	I mm⁴	h mm	I + (A*h²) mm⁴
1			2654	228.0	605112	1340000	102.945	294662244.2
2	12.00	250.00	3000	125.0	375000	15625000	0.555	15625009.1
3	12.00	250.00	3000	125.0	375000	15625000	0.555	15625009.1
4			1310	20.9	27379	572000	104.155	14783255.9
5			1310	20.9	27379	572000	104.155	14783255.9

Area = 11274 mm²   1409870 mm³ I_{NA} = 90282694.1 mm⁴

$\bar{Y}$ = 125.055 mm

To calculate the maximum deflection in the section, from Table 2.2, Case 4:

$$\delta_{max} = \frac{5wL^4}{384EI}$$

$$= \frac{5 \times 13.797 \text{ N/m} \times (6000 \text{ mm})^4}{384 \times 208 \times 10^9 \text{ Pa} \times 90.283 \times 10^{-6} \text{ mm}^4}$$

$$\delta_{max} = 0.012 \text{ mm}$$

The completed physical properties for Example 2.3 are shown in Table 2.6.

Example 2.4

Consider the same uniformly loaded beam section as discussed in Example 2.3, but this time the beam is considered in an encastré condition (Table 2.2, Case 12).

The equation for the calculation of the maximum bending deflection is:

$$\delta_{max} = \frac{wL^4}{384EI} \tag{2.5}$$

Inputting the original values from Example 2.3 gives:

$$\delta_{max} = 2.48 \times 10^{-3} \text{ mm}$$

From this example it can be clearly seen that the maximum deflection of the beam in encastré is significantly less than the beam had it been simply supported.

3 Shaft Design Basics

3.1 INTRODUCTION

Shafts are used in a multitude of engineering examples. There are two basic types of shaft:

- Stationary
- Rotating

Stationary shafts do not carry any torsional loading and are only subject to bending loads such as the king pin as used on the front suspension of an automobile.

Rotating shafts are components of a mechanical device that transmits rotational motion and power. It is integral to any mechanical system where power is being transmitted from a power source, such as an electric motor or an engine to other rotating parts within the system. Examples include gear type speed reducers, belt or chain drives, conveyors, pumps, etc.

Figure 3.1 depicts a typical shaft under discussion. The shaft can carry any number of components including gears, couplings, pulleys, cams, etc. and is generally supported on anti-friction bearings. The torque is transmitted to the shaft mounted components using keys, splines, taper pins, clamping bushes, press fits, etc.

3.2 PROCEDURE FOR DESIGN AND ANALYSIS OF A SHAFT

Figure 3.2 shows a "road map" for the design procedure used in the design of a rotating shaft. The following is a more detailed description of the headings from the road map.

3.2.1 DESIGN REQUIREMENTS FOR THE SHAFT

Before the design can begin it is important to define the end use of the shaft, i.e.:

- To connect a motor to a gearbox or a pump, etc.
- Used in an automotive drive line.
- What will the environment be?
- Will there be any special surface finish requirements?

3.2.2 GEOMETRY OF THE SHAFT

At the initial stage of the design the shaft will have been decided, as well as the mandatory task the shaft will be required to undertake, i.e.:

Transfer power from a drive motor to a gearbox. In this case couplings will be required at each end of the shaft.

Design a turning lathe spindle. Gears will be required to convey the power to the shaft and suitable bearings needed to support the spindle in the machine frame.

Provide a driving force from an axle to a road wheel. This may be in an automotive situation, and here the shaft will be subject to a variable torque as the car or truck accelerates or decelerates during the course of its journey.

These are just some examples of the use of a shaft, and the next phase is to ensure that the shaft is "fit for function" and will not fail during its operating life.

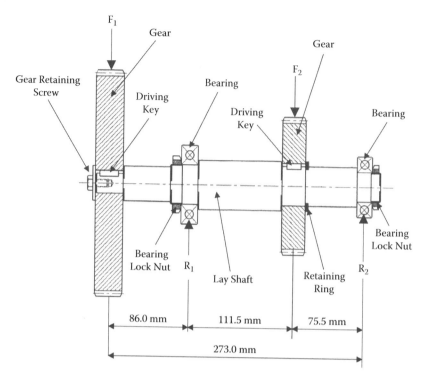

FIGURE 3.1   Typical layshaft.

3.2.3   CALCULATE THE FORCES ACTING ON THE SHAFT

Develop a free-body diagram by replacing the various machine elements mounted on the shaft by their statically equivalent load or torque components.

Specify the location of bearings to support the shaft. The reactions of the bearings supporting radial loads are generally assumed to act at the centre of the bearing. An important point is that in a simple shaft there are only two bearings used to support the shaft, and they should, where possible, be placed on either side of the power transmitting elements to provide a stable support for the shaft and to produce reasonably balanced loading on the bearings. The bearings should be placed as close as possible to the power transmitting elements to minimise bending moments, and in addition, the shaft should be kept as short as possible to keep any shaft deflections at reasonable levels.

There are occasions where due to the length of the shaft (such as in "line shafting"), there may be more than two bearings needed to support the shaft. The analysis of this type of problem is outside the scope of this discussion.

3.2.4   CALCULATE THE BENDING MOMENTS AND SHEAR FORCES ACTING ON THE SHAFT

Example 3.1

Consider Figure 3.1 and construct the bending moment and shear force diagrams for this figure assuming:

$$F_1 = 50 \text{ N}$$

$$F_2 = 20 \text{ N}$$

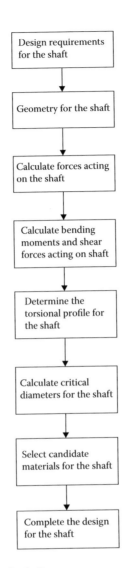

FIGURE 3.2 Road map for the design of a shaft.

Solution:

The first action is to calculate the reactions R_1 and R_2.
Taking moments about R_2:

$$(50.0 \text{ N} \times 273.0 \text{ mm}) + (20.0 \text{ N} \times 75.5 \text{ mm}) = 187.0 \text{ } R_1$$

Therefore R_1 = 81.069 N.
Taking moments about R_1:

$$(50.0 \text{ N} \times 86.0 \text{ mm}) + (187.0 \text{ R2}) = (20.0 \text{ N} \times 111.5 \text{ mm})$$

Therefore R_2 = –11.069 N.
The bending moment acting on the shaft will be calculated next.
From Figure 3.1, the positions of the respective force elements acting on the shaft are next identified and the individual moments then calculated as shown in Table 3.1.

TABLE 3.1

Bending Moments for Example 3.1

Position No.	Calculation	Bending Moment
1	0	= 0
2	(50.0 N × 86.0 mm)	= 4300.0 N.mm
3	(50.0 N × 197.5 mm) − (81.069 N × 111.5 mm)	= 835.806 N.mm
4	(50.0 N × 273.0 mm) − (81.069 N × 187.0 mm) + (20.0 N × 75.5 mm)	= 0.097 N.mm

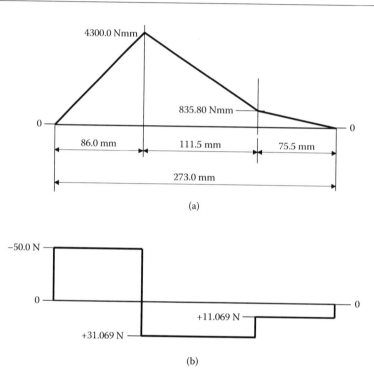

(a)

(b)

FIGURE 3.3 (a) Bending moment diagram for Example 3.1. (b) Shear force diagram for Example 3.1.

The above results are plotted in Figure 3.3(a) and (b); from the results it is obvious that the maximum bending moment occurs at position R_1.

This diagram is drawn for the x:y plane only and would need to be plotted for the y:z plane also if there were forces acting in the y:z plane.

The resultant internal moment at any section along the shaft may be expressed as:

$$M_x = \sqrt{M_{xy}^2 + M_{xz}^2}$$

There may be a number of force inputs into the shaft from gear, chain or pulley drives, and these will introduce variable bending moments in the shaft.

The calculated shear forces will also be considered when finalising the size of the shaft.

3.2.5 DETERMINE THE TORSIONAL PROFILE OF THE SHAFT

If the shaft is connecting a drive motor to, say, a hydraulic pump, in this situation normally the speed of the shaft will be constant. The torque the shaft needs to transmit may be variable depending upon the load being placed on the pump. In general the designer could determine the maximum torque the driving motor can produce and size the shaft based on this calculation, but this may result in a shaft size out of proportion to the duty it will be required to perform and be uneconomical to manufacture. It will be far better to size the shaft based on the maximum load the shaft would see and apply a Safety Factor (SF) to the load.

Questions to ask are: Will the driving load be gradually increased to its operating level? Or, will there be a sharp acceleration from the drive motor followed by a steady load? These factors will greatly influence the torque being applied to the shaft. As an example, in the first case the torsional profile will be a gentle increase followed by a stable period and at the completion of the duty allowed to come gently to rest.

In the second case with the sharply applied acceleration, this will introduce unwanted torsional vibrations in the shaft where the initial torque will be far larger than the shaft would experience when operating normally. In this situation a SF greater than 3 will be needed to ensure a safe life for the shaft.

3.2.6 CALCULATE THE CRITICAL DIAMETERS FOR THE SHAFT

If the shaft is "plain," i.e. no changes to the profile of the shaft, as in a connecting shaft, the calculation will be straightforward, as shown in the following.

Considering a shaft under pure torsion as shown in Figure 3.4.

$$\text{Power} = T\omega \tag{3.1}$$

where
   T = torque in Nm
   ω = angular velocity in rad/s
The unit of power is the watt; W = Nm/s.

This formula can also be written as:

$$\text{Power} = 2\pi nT \tag{3.2}$$

where n = angular velocity in rev/s.

$$\frac{T}{J} = \frac{\tau}{r} = \frac{G\theta}{l} \tag{3.3}$$

where
   T = torque in Nm or Nmm
   J = polar second moment of area in m^4 or mm^4
   r = radius to extreme fibre under stress in m or mm
   G = modulus of rigidity in N/m^2 (Pa)
   τ = maximum shear stress in N/m^2 (Pa)
   l = length of shaft in m or mm

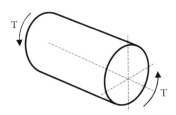

FIGURE 3.4   Basic shaft under torsion.

3.3 SECTION MODULUS

$$Z_t = \frac{J}{r} \tag{3.4}$$

where Z_t is section modulus in m³.

For solid circular sections:

$$T = \frac{\pi}{16} \times d^3 \times \tau \quad \text{or} \quad T = Zt \times \tau \tag{3.5}$$

$$T = \frac{\pi}{16} \frac{D^4 d^4}{D}$$

For hollow sections:

$$T = \frac{\pi}{16} \cdot \left(\frac{D^4 - d^4}{D} \right)$$

where D is the outside diameter in m and d is the inside diameter in m.

For square sections:

$$T = 0.208 \, S^3 \cdot \tau \tag{3.6}$$

where S is the size of the square bar in m.

Values of J and Z_t for various shapes of solid and hollow sections can be found in books covering strength of materials, such as Timoshenko (see Bibliography). Tables 3.5 and 3.6 list properties of some common sections.

Example 3.2

A mild steel shaft is required to transmit 45 kW at 2.5 rev/s.

What diameter of shaft will be suitable using an ultimate shear stress value of 340 MPa? Also calculate the angle of twist over a length of 2.5 m if the modulus of rigidity is 86 GPa.

Solution:

$$\text{Power} = 2 \cdot \pi \cdot n \cdot T$$

$$\text{therefore Torque} \, (T) = \frac{\text{Power}}{2 \cdot \pi \cdot n} = \frac{\pi}{16} d^3 \cdot \tau$$

Allowing for a SF of 4 will give an allowable shear stress (τ) of 85 MPa.

$$d^3 = \frac{\text{Power} \cdot 16}{2\pi^2 \cdot n \cdot t}$$

$$d^3 = \frac{45 \text{ kW} \times 16}{2\pi^2 \times 2.5 \text{ rad/s} \times 85 \text{ MPa}}$$

$$d^3 = 171.65 \times 10^{-6} \text{ m}^3$$

$$d = 0.05556 \text{ m}$$

Say you have a 56 mm diameter; choose a suitable diameter available stock size.

3.3.1 Angle of Twist

$$\frac{T}{J} = \frac{G\theta}{l} = \frac{2\tau}{d}$$

(3.7)

From the standard torsion equation:

$$\theta = \frac{\pi \times d^3 \times \tau \times 32 \times 2.5}{\pi \times d^4 \times 16 \times 86 \times 10^3}$$

$$\theta = \frac{85\,\text{MPa} \times 5\,\text{m}}{0.056\,\text{m} \times 86\,\text{GPa}}$$

Angle of twist in radians = 0.088

Converting to degrees = 0.088 × 57.3 degrees per radian

$\theta = 5.0424$ degrees

Example 3.3

A hollow steel shaft with 220 mm inside diameter and 300 mm outside diameter transmits 2.24 MW at 2 rev/s.
What is the maximum shear stress generated in the shaft material?

Solution:

$$D = 0.30\,\text{m}$$

$$d = 0.22\,\text{m}$$

$$P = 2.24\,\text{MW}$$

$$S = 2\,\text{rev/s}$$

$$T = \frac{2.24\,\text{MW}}{2\pi \times 2}$$

$$T = \frac{\pi}{16}\left(\frac{0.3\,\text{m}^4 - 0.22\,\text{m}^4}{0.3\,\text{m}}\right)\tau$$

Rearranging and solving for τ:

$$\tau = \frac{2.24\,\text{Mw}}{4\pi^2} \times \frac{16 \times 0.3\,\text{m}}{0.3\,\text{m}^4 - 0.22\,\text{m}^3}$$

$$\tau = 47.304\,\text{MPa}$$

This value is compared with the ultimate shear stress value for carbon steel of 480 MPa; hence the shaft is considered to be lightly loaded. (See Table 3.7 for the mechanical properties of a range of common materials.)

3.3.2 ASME Shaft Equations

In 1927 the ANSI/ASME standard for the design of transmission shafting B106-1M-1985 developed the following design equation:

$$D = \left[\frac{32 \cdot N}{\pi} \sqrt{\left[\frac{k_t \cdot M}{S'_n} \right]^2 + \frac{3}{4} \left[\frac{T}{S_y} \right]^2} \right]^{\frac{1}{3}} \tag{3.8}$$

where

k_t = stress concentration factor at shoulder; 1.5 to 2.5

M = bending moment; this is obtained from the bending moment diagram, thus creating a reversed bending moment on the shaft as it rotates (N.m)

T = torsion moment; this is usually uniform (N.m)

N = Factor of Safety (FoS)

D = diameter of shaft at the section being analysed

S'_n = modified endurance strength (this will depend upon the ultimate tensile strength (S_u)); $S'_n = S_n.C_s.C_r$

S_n = endurance strength (MPa)

S_y = torsional yield strength (MPa)

C_s = size factor (Larger diameter shafts tend to have lower fatigue strengths than smaller shafts for a number of reasons. Values of C_s range from about 0.9 for 50 mm diameter shafts to about 0.65 for shafts up to 250 mm diameter.)

C_r = reliability factor (Published fatigue data usually represent an average value of the endurance strength of a sample of test specimens. In the absence of any specific test data the failure distribution is often assumed to follow a normal or Gaussian distribution with a standard deviation of about 8% of the mean. For a 90% nominal reliability C_r is approximately 0.9, and for a 99% reliability C_r is about 0.8; see Table 3.2.)

The ASME equation can only be used subject to the following assumptions:

• Constant torque
• Fully reversed moment
• No axial load applied

Note: This equation can also be derived theoretically from the distortion energy failure theory as applied to fatigue loading.

Example 3.4

It is required to confirm that the critical diameter of a shaft that is manufactured from steel has a minimum diameter of 40.0 mm. The selected shaft material has a fatigue limit of 480 MPa, and the shaft is subject to a bending moment of 340 N.m. The desired reliability factor is 0.99.

TABLE 3.2
Reliability Factors

Desired Reliability	Reliability Factor (Cr)
0.50	1.00
0.90	0.90
0.99	0.81
0.999	0.75

A shoulder with a fillet radius of 0.50 mm is provided on the shaft to locate a bearing; a FoS of 2.5 is required.

Solution:

The torque carried by the shaft:

$$\omega = 2\pi n$$

$$\omega = 2 \times p \times 4 \text{ rad/s}$$

$$= 25.13274 \text{ rad/s}$$

$$Torque = \frac{power}{\omega}$$

$$= 178 \text{ N.m}$$

For Equation (3.8) the following values are to be inputs:

$$N = 2.5 \text{ (safety factor)}$$

$$k_t = 1.75$$

$$M = 340 \text{ N.m}$$

$$T = 178 \text{ N.m}$$

$$S_n = 480 \text{ MPa}$$

$$C_s = 0.9$$

$$C_r = 0.81$$

$$S'_n = 349.92$$

$$S_y = 125 \text{ MPa}$$

$$D = \left[\left[\frac{32 \times 2.5}{\pi} \right] \times \sqrt{ \left[\frac{1.75 \times 340 \text{ N} \cdot \text{m}}{349.92 \text{ MPa}} \right]^2 + \frac{3}{4} \times \left[\frac{178 \text{ N} \cdot \text{m}}{125 \text{ MPa}} \right]^2 } \right]^{\frac{1}{3}}$$

$$D = 37.678 \text{ mm}$$

From the above calculation it is considered that the critical diameter of 40.0 mm is satisfactory.

3.3.3 Fillet Radii and Stress Concentrations

When there is a change in the diameter of a shaft to create a shoulder against which to locate a machine element such as a gear or bearing, depending on the ratio of the two diameters and the radius in the fillet, a stress concentration will be generated. It is recommended that the fillet radius r be as large as possible to minimise the stress concentration. At times if the part is proprietary,

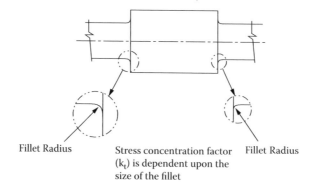

Fillet Radius   Stress concentration factor   Fillet Radius
                   (k_t) is dependent upon the
                   size of the fillet

FIGURE 3.5   Fillets on a shaft.

the designer has no control over the radius and has to accommodate it in the design. For the purpose of design, fillets are classified into two categories: sharp and well rounded (see Figure 3.5).

Although the term *sharp* is used, this does not mean truly sharp and without any fillet radius at all. Such a shoulder configuration would have a very high stress concentration factor and should therefore be avoided. One such situation where it is likely to occur is where a ball or roller bearing is to be located. The inner race of the bearing has a factory produced radius, but it is small; therefore the fillet radius on the shaft will have to be smaller to allow the bearing to sit against the shoulder. Where the element has a large chamfer on its bore and is located against a shoulder, or where there is nothing locating against the shoulder, the fillet radius can be as large (well rounded) as possible and the corresponding stress concentration is smaller.

The symbol used for stress concentration is k.

The following values are used for bending:

$$k_b = 2.5 \text{ (sharp fillet)}$$

$$k_b = 1.5 \text{ (well rounded fillet)}$$

It is possible to establish the stress concentration factor for a specific application. The following formula is used for elastic bending stress:

$$k = K_1 + K_2\left[\frac{2h}{D}\right] + K_3\left(\frac{2h}{D}\right)^2 + K_4\left(\frac{2h}{D}\right)^3 ..... \tag{3.9}$$

where
  h = D − d
  D = large diameter on shaft
  d = adjacent smaller diameter of shaft

Table 3.3 gives the coefficients for bending stress concentration factors, and there is a similar table covering elastic torsional stress which is shown in Table 3.4.

3.3.4   Undercuts

Most ball and roller bearings are manufactured with small fillet radii in their bores, and in these situations where the diameter of the shaft is stress critical, particularly in the vicinity of a bearing, it is possible to provide a local undercut in the shaft adjacent to the shoulder. Figure 3.6 provides details of such an undercut. This allows the bearing to be fitted against the abutment

TABLE 3.3
Coefficients for Bending Stress Concentration Factors

	$0.25 \le \dfrac{h}{r} \le 2.0$	$2.0 \le \dfrac{h}{r} \le 20.0$
K_1	$0.927 + 1.149\sqrt{h/r} - 0.08\ h/r$	$1.225 + 0.831\sqrt{h/r} - 0.010\ h/r$
K_2	$0.015 - 3.281\sqrt{h/r} + 0.837\ h/r$	$-3.790 + 0.958\sqrt{h/r} - 0.257\ h/r$
K_3	$0.847 + 1.716\sqrt{h/r} - 0.506\ h/r$	$7.374 - 4.834\sqrt{h/r} + 0.862\ h/r$
K_4	$-0.790 + 0.417\sqrt{h/r} - 0.246\ h/r$	$-3.809 + 3.046\sqrt{h/r} - 0.595\ h/r$

TABLE 3.4
Coefficients for Elastic Stress Concentration Factors from Table 3.3

	$0.25 \le \dfrac{h}{r} \le 4.0$
K_1	$-1.081 + 0.232\sqrt{h/r} + 0.065\ h/r$
K_2	$-0.493 - 1.820\sqrt{h/r} + 0.517\ h/r$
K_3	$1.621 + 0.908\sqrt{h/r} - 0.529\ h/r$
K_4	$-1.081 + 0.232\sqrt{h/r} + 0.065\ h/r$

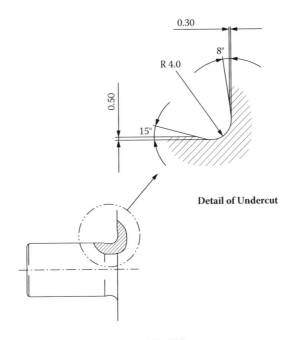

FIGURE 3.6 Detail of undercut in a shaft based on DIN 509.

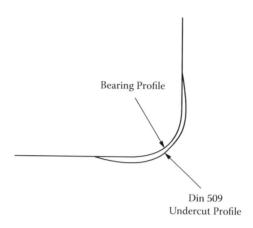

Bearing Profile

Din 509
Undercut Profile

FIGURE 3.7   Comparison between bearing and undercut profiles.

face and still provide a reasonable fillet radius. The undercut will allow the stress concentration local to the fillet radius to be reduced. Care will need to be exercised to ensure that the presence of the undercut does not compromise the local strength of the shaft as the diameter is reduced a little, and that the revised SF is still acceptable. Figure 3.7 shows a comparison between a roller bearing fillet radius and the undercut fillet radius, and it can be seen that the undercut allows a slightly larger fillet radius to be made, thereby reducing the stress concentration factor (k_t) by a small amount.

1. Establish the most efficient configuration for the shaft. Position the components along the shaft.
2. Select the most appropriate method for driving the shaft.
   Determine load profile
   Reactions/shear stresses/bending moments
   Dimensions of shaft at critical points
3. Select the most appropriate bearings and components. The following rules are proposed to enable a suitable shaft to be derived to meet the design requirement.
   a. Develop a free-body diagram by replacing the various machine elements mounted on the shaft by their statically equivalent load or torque components. Specify the location of bearings to support the shaft. The reactions of the bearings supporting radial loads are assumed to act at the centre of the bearing. An important point is that generally there are only two bearings used to support the shaft, and they should, where possible, be placed on either side of the power transmitting elements to provide a stable support for the shaft and to produce reasonably balanced loading on the bearings. The bearings should be placed as close as possible to the power transmitting elements to minimise bending moments, and in addition, the shaft should be kept as short as possible to keep any shaft deflections at reasonable levels.
   b. Draw a bending moment diagram in the x-y and x-z planes as shown in Figure 3.3(a) and (b). The resultant internal moment at any section along the shaft may be expressed as:

$$M_x = \sqrt{M_{xy}^2 + M_{xz}^2}$$

   c. Determine the magnitude of the power and torque to be transmitted by the shaft.

d. Develop a torque diagram. Torque developed from one power transmitting element must balance the torque from other power transmitting elements.

e. Establish the location of the critical cross section, or the x location where the torque and moment are the largest.

f. Select the material which the shaft will be manufactured from, specifying:
   - Ultimate tensile strength (σ_u)
   - Yield strength (σ_y)
   - Surface condition:
      Machined
      Ground
      Hot rolled

g. Consider the method of how each element on the shaft will be held in position axially and how the power transmission from each element to the shaft will take place. Design details such as fillet radii, shoulder heights and key seat dimensions should be specified.

h. Determine the forces being exerted on the shaft.

i. Resolve the radial forces into orthogonal components (vertical and horizontal forces).

j. Resolve the final reactions on the end supports (bearings) in each plane.

k. Analyse each critical point on the shaft to determine the minimum acceptable diameter to ensure adequate SFs for the loading at that point on the shaft.

   In general these critical points will include where a change in diameter will take place, where higher values of torque and bending moments will occur, together with stress concentrations that also occur.

l. Due to the nature of the shaft, as the shaft rotates, tensile and compression stresses will occur on either side of the shaft at each rotation. Carry out an endurance study to ensure the life of the shaft will not be compromised due to fatigue.

TABLE 3.5
Properties for Some Common Sections

	Circular Sections		Non-Circular Sections		
	Solid	**Tubular**	**Square**	**Rect'**	**Thin tube**
	D = outer dia'	d = inner dia'	D (side), D (side)	D (height), B (width)	t = thickness at any point U = length of median
A	$\dfrac{\pi}{4} \cdot D^4$	$\dfrac{\pi}{4} \cdot (D^2 - d^2)$	D^2	BD	Mean of areas enclosed by outer and inner boundaries
I_p	$\dfrac{\pi}{32} \cdot D^4$	$\dfrac{\pi}{32} \cdot (D^4 - d^4)$	$0.1406\, D^4$	βBD^3	$\dfrac{4A^2}{\int (dU/t)}$
Z_p	$\dfrac{\pi}{16} \cdot D^3$	$\dfrac{\pi}{16 \cdot D} \cdot (D^4 - d^4)$	$0.2082\, D^3$	$\alpha\beta D^2$	$2\,At$ t = minimum value

K (for square): $0.1406\, D^4$; K (for rect'): βBD^3

TABLE 3.6

Coefficients for Evaluating Rectangular Sections from Table 3.5

B/D	1.0	1.5	2.0	2.5	3.0	4.0	6.0	8.0	≥10.0
α	0.208	0.231	0.246	0.258	0.267	0.282	0.299	0.307	0.333
β	0.141	0.196	0.229	0.249	0.263	0.281	0.299	0.307	0.333

TABLE 3.7

Mechanical Properties for a Range of Common Materials

Mechanical Properties	Carbon Steel	Stainless Steel	Aluminium Alloy	Phosphor Bronze
Specification	220M07 (EN 1A)	304	BS 2014 T6 condition	BS 2874 PB-102
Ultimate strength (f_t)	480 N/mm²	515 N/mm² (min.)	483 N/mm²	280 N/mm²
Yield strength (0.2%) (t_2)	280 N/mm²	205 N/mm² (min.)	414 N/mm²	120 N/mm²
Shear strength (f_{so})	727 N/mm² (estimated)	340 N/mm² (estimated min.)	290 M/mm²	176 N/mm² (estimated)
Poisson's ratio (υ)	0.28	0.30	0.33	0.355
Elongation (%)	14	40	8	15
Modulus of elasticity (E)	195 kN/mm²	193	73.1 kN/mm²	110 kN/mm²
Shear modulus (G)	77 kN/mm²	82	27 kN/mm²	41 kN/mm²
Density (ρ)	7800 kg/m³	8000 kg/m³	2800 kg/m²	8000 kg/m³

4 Combined Torsion and Bending

Very often bending and torsion do not act alone when designing a component; they can often act simultaneously. As an example, a shaft that is supported as a cantilever with a drive pulley fitted on the free end will in this case be resisting a bending load due to the drive belt or chain tension and a torsion load again imposed due to the drive belt or chain. This example is not restricted to rotating shafts but can equally be applied to brackets or other nonrotating components.

In the first instance the bending moment (M) needs to be calculated and then the twisting moment (T) requires evaluation.

From the bending equation:

$$\frac{M}{I} = \frac{\sigma}{y} = \frac{E}{R} \tag{4.1}$$

from which:

$$\sigma = \frac{M.y}{I} \tag{4.2}$$

and

$$\sigma = \frac{E.y}{R} \tag{4.3}$$

The torsion equation:

$$\frac{T}{J} = \frac{\tau}{r} = \frac{G\theta}{L} \tag{4.4}$$

from which:

$$\tau = \frac{T.r}{J} \tag{4.5}$$

and

$$\theta = \frac{\tau.L}{G.r} \tag{4.6}$$

For notation see Table 4.1.

When designing shafts or components subjected to combined torsion and bending for maximum strength, the material selected will determine the nature of the failure.

$$\text{i.e. when} \quad \frac{\sigma_x - \sigma_y}{2} = \frac{\sigma_E}{2} \tag{4.7}$$

$$\text{or} \quad \sigma_x - \sigma_y = \sigma_E \tag{4.8}$$

TABLE 4.1

Notation

M	Maximum bending moment	Nm
I	Moment of inertia (second moment of area)	m^4
σ	Extreme fibre stress due to bending	Pa
y	Distance from the neutral axis to extreme fibre	m
E	Modulus of elasticity	N/m²
R	Radius of bending on neutral axis	m
T	Twisting moment or torque	Nm
J	Polar moment of inertia	m^4
τ	Maximum shear stress due to twisting	Pa
r	Radius of circular section	m
G	Modulus of rigidity	N/m²
θ	Angle of twist over length	Radians
L	Length	m
M_E	Equivalent bending moment due to combined bending and torque	Nm
T_E	Equivalent torque due to combined moment and torque	Nm

For *ductile materials* such as mild steel, failure is considered to occur when the greatest shear stress reaches the maximum shear stress at the elastic limit in a simple tension test.

This is known as the *maximum-shear-stress theory* (*Guest's* or *Tresca criterion*) and gives good correlation with experimental results obtained with ductile materials.

It can be shown:

$$\tau = \pm\sqrt{(\sigma^2 + 4\tau^2)} \tag{4.9}$$

and the equivalent torque

$$T_E = \sqrt{(M^2 + T^2)}$$
$$= \frac{\pi \times d^2}{16} \times \tau \tag{4.10}$$

When dealing with *brittle materials* such as cast iron that is subjected to combined torsion and bending, it is more important to determine the value of the principal stress, which in this case is a tensile stress and has a greater effect than shear stress. This approach is known as *Rankine's theory* (brittle materials will be generally weaker in tension than shear).

Rankine's theory for brittle materials is:

$$\sigma = \frac{\sigma}{2} \pm \frac{1}{2}\sqrt{(\sigma^2 + 4\tau^2)} \tag{4.11}$$

The equivalent bending moment:

$$M_E = \frac{1}{2}\left[M + \sqrt{(M^2 + T^2)}\right] \tag{4.12}$$

Example 4.1

Part of a spur reduction gear of 5:1 is shown in Figure 4.1. The pinion has 20 teeth with a module of 6 mm and an involute angle of 20°. The power transmitted is 45 kW at 13.5 rev/s.

Calculate the diameter of the pinion and wheel shafts taking the allowable shear stress as 55 MPa.

Solution:

$$\text{Pitch diameter} = \text{module} \times \text{number of teeth}$$

$$= 6 \text{ module} \times 20 \text{ teeth}$$

$$= 120 \text{ mm}$$

and

$$\text{Torque} = \frac{\text{power}}{2\pi N}$$

Tangential tooth load (E):

$$E = \frac{\text{Torque}}{\text{radius of pitch circle}}$$

$$= 530.44 \text{ Nm}$$

$$= 8840.67 \text{ N}$$

Involute angle $\alpha = 20°$
Maximum tooth load:

$$P = \frac{E}{\text{Cos}20°}$$

$$= \frac{8840.67}{0.9397}$$

$$= 9.408 \text{ kN}$$

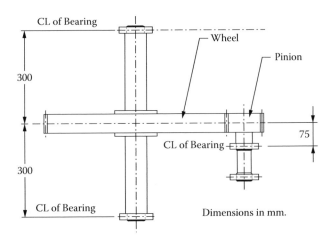

FIGURE 4.1 Spur reduction gear.

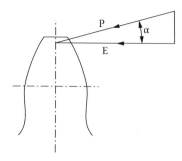

FIGURE 4.2   Forces acting on gear tooth.

The force E is the driving force and puts a torque on the shafts while P produces a bending effect (see Figure 4.2).

For the pinion shaft:

$$\text{Torsional moment T} = 530.44 \text{ Nm}$$

$$\text{Bending moment M} = 9408 \text{ N} \times 0.075 \text{ m}$$

$$\text{Equivalent torque } T_E = \sqrt{M^2 + T^2}$$

$$= \sqrt{705.6^2 + 530.44^2}$$

$$= \sqrt{779237}$$

$$= 705.6 \text{ Nm}$$

$$= 882.74 \text{ Nm}$$

Rearranging for d^3 and the allowable shear stress $\tau = 55$ MPa.

$$882.74\,\text{Nm} = \frac{\pi}{16} d^3 \times 55 \times 10^6$$

$$d^3 = \frac{882.74 \times 16}{\pi \times 55 \times 10^6}$$

$$d = 43.4 \text{ mm, say.}$$

For the wheel shaft:

$$\text{Gear ratio} = 5{:}1$$

$$\text{Torsional moment} = 530.44 \text{ Nm} \times 5$$

$$= 2652.2 \text{ Nm}$$

$$\textit{Bending moment} = \frac{WL}{4}$$

$$= 9408 \text{ N} \times \frac{0.6}{4} \text{ m}$$

$$= 2652.2 \text{ Nm}$$

$$= 1411.2 \text{ Nm}$$

The reactions are assumed to act at the centre of the bearing.

$$= \sqrt{1411.2^2 + 2652.2^2}$$

Equivalent shear stress T_E or torque $= 3004.27$ Nm

$$d^3 = 0.000278$$

$$d = 0.0653 \text{ m}$$

$$d = 65.3 \text{ mm}$$

If an allowance is made for the keyway:

$d = 65.3 \times 1.1$ (A factor of 1.1 is included to allow for the reduction in strength due to the presence of the keyway.)

$$d = 71.83 \text{ mm}$$

Example 4.2

Figure 4.3 shows a cast steel bracket subject to a force of 9 kN in the given directions. Calculate:

- The stress due to bending across the plane y:y.
- The stress due to the combined bending and torsion across the plane x:x.
- The stresses in the flange bolts resisting the torsional, sliding and overturning actions of the load.
- State the nature of the stresses in each case.

Solution:

Considering section y:y

$$M = \sigma Z \quad \left(\text{where } Z = \frac{\pi}{32^2}d^3 \text{ for circular section}\right)$$

$$9 \times 0.28 \text{ kN} = \frac{\pi \times 0.100^3}{32} \times \sigma$$

$$\sigma = \frac{9 \times 0.28 \times 32}{\pi \times 0.100^3}$$

$$= 25668.5 \text{ kN/m}^2$$

$$= 25.67 \text{ MPa}$$

The bending moment (M) acting on this section is 9 kN × 280 mm.

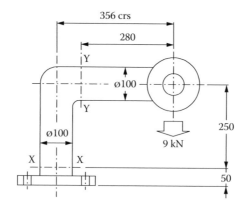

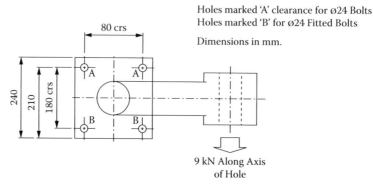

Holes marked 'A' clearance for ø24 Bolts
Holes marked 'B' for ø24 Fitted Bolts

Dimensions in mm.

9 kN Along Axis
of Hole

FIGURE 4.3   Steel bracket in Example 4.2.

Considering section x:x

$$M = 9 \text{ kN} \times 250 \text{ mm}$$

$$= 2.25 \text{ kNm}$$

The bending moment (M) acting on this section is 9 kN × 250 mm.
Torsional moment (T):

$$T = 9 \text{ kN} \times 356 \text{ mm}$$

$$= 3.204 \text{ kNm}$$

$$Te = \sqrt{m^2 + T^2}$$

$$Te = \sqrt{2.25^2 + 3.204^2}$$

$$Te = 3.915 \text{ kNm}$$

$$T_{max} = \frac{16}{\pi d^3}\sqrt{m^2 + T^2}$$

$$T_{max} = \frac{16}{\pi \times 0.1^3 \text{m}^3}\sqrt{2.25^2 + 3.204^2}$$

$$T_{max} = \frac{16}{\pi \times 0.1^3 m^3} \times 3.915 \text{ kNm}$$

$$T_{max} = 19938.9 \text{ kN/m}^2$$

$$T_{max} = 19.9389 \text{ MPa}$$

This is the maximum shear stress across section x:x due to the combined torsion and bending.

Consider the stresses acting on the bolts:
The twisting and sliding will generate a shear stress in the fasteners.
Consider the two fitted bolts only, as these will resist the sliding forces (see Figure 4.4).

$$= 20.1^2 \times \frac{\pi}{4}$$

$$= 317.310 \text{ mm}^2$$

Shear stress on bolt Y:

$$= \frac{13.3 \text{ kN}}{317.31 \text{mm}^2}$$

$$= 0.0419 \text{ kN/mm}^2$$

$$= 0.0419 \text{ MPa}$$

Figure 4.5 shows the reactions acting on the bolts due to the load of 9 kN.

$$= \frac{22.3 \text{ kN}}{317.31 \text{ mm}^2}$$

$$= 0.0703 \text{ kN/mm}^2$$

$$= 0.0703 \text{ MPa}$$

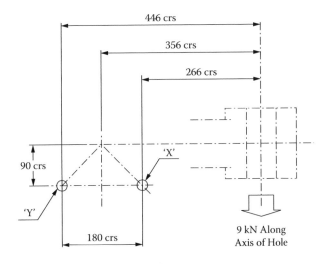

FIGURE 4.4 Bolts subject to shear load.

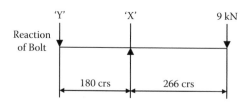

FIGURE 4.5   Reactions at fasteners.

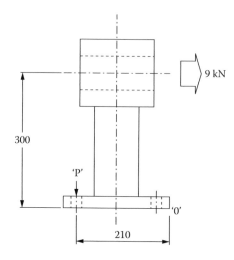

FIGURE 4.6   Overturning forces acting on fasteners.

Now the root area of Ø24 mm diameter bolt:

$$= 317.31 \text{ mm}^2$$

The two bolts with clearance holes will resist the overturning effect about the edge O in Figure 4.6.

Shear stress acting on the bolt:

$$\text{Overturning moment} = 9 \text{ kN} \times 300 \text{ mm}$$

$$\text{Resisting moment} = P \times 210 \text{ mm}$$

$$9 \times 300 = P \times 210$$

$$P = \frac{9 \times 300}{210} \text{ kN}$$

$$P = 12.86 \text{ kN}$$

$$\text{Now load per bolt} = \frac{12.86}{2} \text{ kN}$$

$$= 6.43 \text{ kN}$$

$$\text{Area at Root of } \varphi 24 = 317.31 \text{ mm}^2$$

$$\text{Tensile stress in bolts } = \frac{6.43 \text{ kN}}{317.31 \text{ mm}^2}$$

$$= 20.3 \text{ MPa}$$

Example 4.3

The overhead gear for an electric motorised lift is shown in Figure 4.7. The sheave is mounted and keyed centrally on the shaft, which is supported on bearing 510 mm between centres.

The ropes pass from the edge of the cage over the sheaves to the balancing mass.

The mass of the cage, the mass carried by the cage and the balancing mass are 1100, 910 and 1550 kg, respectively.

The maximum acceleration of the cage is 2 m/s^2, and assuming the shaft to be simply supported at the bearings, find the diameter of the shaft, using a maximum allowable shear stress of 40 MPa (neglect the inertia effect of the sheave, and friction in the bearings is negligible).

Solution:

It is possible to apply shock and fatigue factors to the bending and twisting moments; values for these factors can be based on those given in Table 4.2, depending on the nature of the loading.

Tension in cable supporting the ascending cage:
To consider the tension in the cable, the dynamic situation is imagined to be static by applying d'Alembert's principle, which is stated briefly by putting an inertia force of m × a in the opposite direction to the acceleration.

Tension in cable (T_1).
Gravitational force (mg).
Inertia force (ma).

See Figure 4.8(a).

$$T_1 = mg + ma$$

$$T_1 = (1100 + 910) \text{ kg} \times 9.81 \text{ m/s}^2 + (1100 + 910) \times 2 \text{ kg.m/s}^2$$

$$T_1 = 19.72 \text{ kN} + 4.02 \text{ kN}$$

$$T_1 = 23.74 \text{ kN}$$

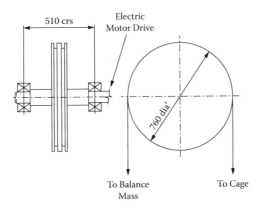

FIGURE 4.7 Overhead gear in Example 4.3.

TABLE 4.2

Factors of Safety for Various Applications

Stationary				Rotating		
	Factor k_b Bending	Factor k_t Torsion			Factor k_b Bending	Factor k_t Torsion
Load applied gradually	1.0	1.0	Gradually		1.5	1.0
Load applied suddenly	1.5–2.0	1.5–2.0	Minor sudden shock		1.5–2.0	1.0–1.5
			Sudden heavy shock		2.0–3.0	1.5–3.0

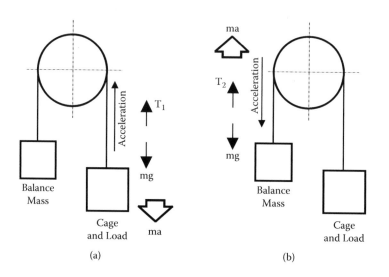

FIGURE 4.8   (a) Tension in cable due to ascending cage. (b) Tension in cable due to descending cage.

Tension in cable supporting descending balance mass:
See Figure 4.8(b).

$$T_2 + ma = mg$$

$$T_2 = mg - ma$$

$$= 1550 \times 9.81 - 1550 \times 2$$

$$= 15.2 - 3.1 \text{ kN}$$

$$T_2 = 12.1 \text{ kN}$$

$$\text{Torque} = (T_1 - T_2) \times \text{radius of sheave}$$

$$T = (23.74 - 12.1) \times \frac{760}{2} \times 10^{-3} \text{ kNm}$$

$$T = 11.64 \times 0.38 \text{ kNm}$$

$$T = 4.423 \text{ kNm}$$

Torque supplied by motor.

Bending of shaft:
Total bending force:

$$= 23.74 + 12.1 \text{ kN}$$

$$= 35.84 \text{ kN}$$

$$= \frac{F}{2} \times \frac{L}{2}$$

$$= \frac{35.83}{2} \times \frac{0.510}{2}$$

Maximum bending moment:

$$BM_{max} = 4.57 \text{ kNm}$$

To find the equivalent torsional moment and hence shaft diameter, consider the load as being applied as a sudden minor shock with a factor $(k_b) = 1.5$ and a factor $(k_t) = 1.1$.

$$d^3 = \frac{16}{\pi \times t} \times \sqrt{((k_b M)^2 + (k_t T)^2)}$$

$$d^3 = \frac{16}{\pi \times 40 \times 10^3} \times ((1.5 \times 4.57)^2 + (1.1 \times 4.423)^2)$$

$$d^3 = 0.0001273\sqrt{(47 + 23.7)}$$

$$= 0.0010738$$

$$d = 0.1022 \text{ m}$$

$$d = 102.2 \text{ mm}$$

Making an allowance for a keyway (as previously stated),

$$102.2 \text{ mm} \times 1.1 = 112.4 \text{ mm}$$

Now

$$\sigma_t = \frac{M.y}{I}$$

where
M = 4.57 × 10³ Nm
y = d/2 = 53.5 mm

$$I = \frac{\pi.D^4}{32}$$

$$I = \frac{\pi \times (107.31 \text{ mm})^4}{32}$$

$$= 13.018 \times 10^{-6}$$

$$\sigma_t = \frac{4.57 \times 10^3 \text{ Nm} \times 57.50 \text{ mm}}{13.018 \times 10^{-6} \text{ mm}^4}$$

$$= 18.78 \times 10^6 \text{ Pa}$$

It is suggested that the shaft may be manufactured from mild steel 220M07 (EN1A) where $\sigma_t = 480 \times 10^6$ Pa will give a Safety Factor (SF) > 6. Hence the shaft will be reasonably resistant to fatigue.

For a suitable stock size select a 110.0 mm diameter.

5 Keys and Spline Calculations

5.1 INTRODUCTION

Keys and splines allow easy attachment and detachment of couplings and gears to shafts and are found in any number of examples in engineering where this feature is required. The types of keyways and splines shown in Figure 5.1 are considered in this chapter and include the following.

5.1.1 FEATHER KEY

The feather key is considered the least strong connection and is restricted to the simpler connections where load reversals are minimal and where the shear strength of the key is considered sufficient.

5.1.2 STRAIGHT SPLINE

The straight sided spline is able to carry reversing loads and is reasonably simple to manufacture. The load carrying capacity of the spline is obviously better than the simple feather key, but due to pitch errors, not all the splines carry the full load, allowing other splines to become overloaded; therefore considerable care is required in the calculation to ensure there is adequate strength. The manufacture of the spline can be facilitated using fairly simple workshop machine tools using an indexing fixture. Most straight sided splines generally have six teeth, but this can be increased to eight teeth.

5.1.3 INVOLUTE SPLINE

The involute spline is judged to be the strongest connection method as the number of teeth is increased over that of the straight spline. Up to 24 teeth on the larger diameter splines is reasonable where the load carrying capacity is very high. They are also able to take high reversing loads. The manufacture is restricted to gear hobbing machines, which limits the types of manufacturing facilities to manufacture them.

Because the tooth form is essentially based on the gear tooth, pitch errors are minimised, ensuring a far greater number of teeth able to carry the torsional load. In any shaft calculation the strength of the shaft will dictate the load carrying capacity, and therefore this will be the first requirement to be calculated.

5.2 PROCEDURE FOR ESTIMATING THE STRENGTH CAPACITY OF SHAFT

1. Estimate the capacity of the shaft for transmitting the required torque with a reduced shaft diameter, that is the diameter of the shaft that does not include either the depth of the key or spline (see Figure 5.2). The general nomenclature for a key or spline is covered in Table 5.1. The nomenclature for a key connection is covered in Table 5.2, and for a straight sided spline, see Table 5.3.
2. Evaluate the shear and compressive stresses in the shaft and either the single key or spline.
3. Consider the service factor relating to the application factor, design factor, fatigue life factor and wear factor from the tables for the particular design and application being considered.

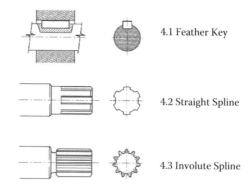

4.1 Feather Key

4.2 Straight Spline

4.3 Involute Spline

FIGURE 5.1   Types of connections covered in this chapter.

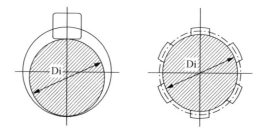

FIGURE 5.2   Reduced diameter for keys and splines.

TABLE 5.1

Nomenclature for Key and Spline

D_o	= Shaft outside diameter	m
D_i	= Reduced diameter	m
D_h	= Diameter of hole – hollow shaft	
T	= Applied torque	Nm
K_s	= Service factor	
σ_c	= Compressive stress of key	N/m²
τ	= Shear stress in key	N/m²

Shear stress (τ) generated in a solid shaft due to an applied torque:

$$\tau = \frac{16 \cdot T \cdot K_s}{\pi \cdot D_i^3} \tag{5.1}$$

Shear stress (τ) generated in a hollow shaft due to an applied torque:

$$\tau = \frac{16 \cdot T \cdot D_i \cdot K_s}{\pi \cdot \left(D_i^4 - D_h^4\right)} \tag{5.2}$$

5.3 STRENGTH CAPACITY OF KEY

$$x = t_1 - \left(r - \sqrt{r^2 - (W/2)^2}\right) - c \tag{5.3}$$

TABLE 5.2
Nomenclature for Key

D	= Nominal diameter of shaft	m
x	= Depth of keyface (key/shaft) taking side force	m
x_1	= Depth of keyface (key/hub) taking side force	m
r	= Radius of shaft ($D_o/2$)	m
L_e	= Effective length of key	m
c	= Key chamfer size	m
d	= Depth of key	m
w	= Width of key	m
t_1	= Depth of keyway	m
T	= Applied torque	Nm
F	= Force acting on key (T/r)	N
K_s	= Service factor	
	For fixed/close fit keyway, $K_s = K_a \cdot K_d / K_f$	
	For sliding fit keyway, $K_s = K_a \cdot K_d / K_w$	
σ_c	= Resulting compressive stress on key	N/m²
τ	= Resulting shear stress in key	N/m²

TABLE 5.3
Nomenclature for Straight Sided Spline

D_m	= Mean (pitch) diameter	m
r	= Mean radius of spline ($D_m/2$)	m
n	= Number of splines	
L_e	= Effective length of spline (straight length)	m
d	= Depth of spline	m
w	= Width of spline	m
T	= Applied torque	Nm
K_s	= Service factor	
	For a fixed/guided spline, $K_s = K_a \cdot K_d / K_f$	
	For a flexible/sliding fit spline, $K_s = K_a \cdot K_m \cdot K_d / K_w$	
σ_c	= Resulting compressive stress in spline	Nm²
τ	= Resulting shear stress in spline	Nm²

$$x_1 = D - x - 2 \cdot c$$

For a 25.000 mm diameter shaft fitted with a 8.0 mm × 7.0 mm key,

$$x = 4 - \left(12.5 - \sqrt{12.2 - 42}\right) - 0.4 \tag{5.4}$$

$$= 2.94 \text{ mm}$$

$$x_1 = 7 - 2.94$$

$$= 3.26 \text{ mm}$$

where

K_s = Service factor
K_a = Application factor

K_d = Design factor
K_f = Fatigue life factor
K_w = Wear life factor
For values of these factors, see Tables 5.9 to 5.13.

Note: It is considered reasonable to use x = D/2; however, Figure 5.3 will provide a more accurate value of x.

5.4   STRENGTH CAPACITY OF AN ISO STRAIGHT SIDED SPLINE

For the nomenclature used for an ISO straight sided spline see Figure 5.4 and Table 5.3.
    Compressive strength of straight sided spline:

$$\sigma = \frac{T \cdot K_s}{n \cdot d \cdot L_e \cdot r} \tag{5.5}$$

5.5   STRENGTH CAPACITY OF ISO INVOLUTE SPLINE

For the nomenclature used for involute splines see Figure 5.5 and Table 5.4.

Note:

For 30° flat root splines, h = 0.9 m.
For fillet root splines, h = m.

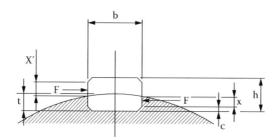

FIGURE 5.3   Nomenclature for key.

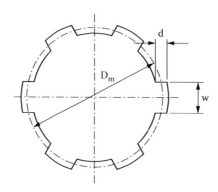

FIGURE 5.4   Nomenclature for straight sided spline.

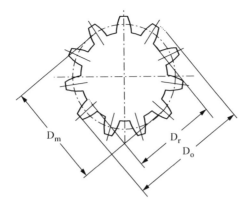

FIGURE 5.5 Nomenclature for involute spline.

TABLE 5.4

Nomenclature for Involute Splines

D_m	= Mean (pitch) diameter	m
D_o	= Outside diameter	m
D_r	= Root diameter	m
m	= Module	m
t	= Tooth thickness	m
h	= Depth of engagement of spline teeth	m
n	= No. of teeth	
L_e	= Length of engagement (straight length)	m
T	= Applied torque	Nm
K_s	= Service factor	
	For a fixed close fit/guided spline, $K_s = K_a/K_f$	
	For a flexible/sliding spline, $K_s = K_m.K_d/k_w$	
σ_c	= Resulting compressive stress in shaft	N/m^2
τ	= Resulting shear stress in shaft	N/m^2

Shear stress at mean diameter of spline resulting from the applied torque, including a factor of 2 (assuming only half the teeth are loaded due to pitch errors):

$$\tau = \frac{4 \cdot T \cdot K_s}{D_m \cdot n \cdot L_e \cdot h} \tag{5.6}$$

The compressive stress in teeth due to the applied torque:

$$\sigma_c = \frac{2 \cdot T \cdot K_s}{D_m \cdot n \cdot L_e \cdot h} \tag{5.7}$$

Tables 5.9, 5.10, 5.11, 5.12 and 5.13 give the respective values for the factors K_d, K_a, K_m, K_f and K_w used in the calculations.

5.6   EXAMPLE CALCULATIONS

The following example calculations compare the strength of a keyed shaft, a straight sided spline and an involute spline and are based on identical transmitted power and rev/s. The keys and splines are all considered to be close fitting and guided.

For input data for examples, see Table 5.5.
For details of factors used in this example, see Table 5.6.
Now:

$$P = \frac{T\omega}{9549}$$

where
  P is in kW
  T is in Nm
  ω is in rev/min
  Solving for torque:

$$T = \frac{9549 \times P}{\omega}$$

$$= \frac{9549 \times 10 \text{ kW}}{1500 \text{ rev/min}}$$

$$T = 63.6 \text{ Nm}$$

TABLE 5.5

Input Data for Example Calculations

1	Power transferred	10	kW
2	Rotational speed	25	rev/s
3	Shaft diameter	25	mm
4	Resulting torque (P/n.9549)	63.66	Nm
5	Design life	10×10^3	h
6	Number of start/stops	10×10^4	h
7	Total no. of revolutions of shaft	900×10^6	rev
8	Materials of construction:		
	BS EN 10083 C45 normalised		
	σ_t—ultimate tensile strength	580	MPa
	σ_y—yield strength	300	MPa
	σ_s—shear strength	200	MPa
	σ_c—compressive strength	130	MPa

TABLE 5.6

Factors Used for Example Calculations

K_d	Design factor	1.0
K_a	Application factor	1.0
K_m	Load distribution factor	1.0
K_f	Fatigue factor	0.5
K_w	Wear factor	0.7
K_s	Service factor:	
K_{shaft}	Shaft	2.0
K_k	Key (close fit)	2.0
K_{sp}	Spline (close fit)	2.0

5.6.1 Shaft Calculations

Assume a basic shaft diameter of 25.00 mm.

Shear stress in shaft (τ):

$$\tau = \frac{16 \cdot T \cdot K_s}{\pi \cdot D_o^3} \tag{5.8}$$

$$= \frac{16 \times 63.66 \text{ Nm} \times 2}{\pi \times (25.00 \text{ mm})^3}$$

$$= 41.5 \text{ MPa}$$

Factor of Safety (FoS):

$$\text{FoS} = \frac{\sigma_s}{\tau} \tag{5.9}$$

$$= \frac{200 \text{ MPa}}{41.5 \text{ MPa}}$$

$$\text{FoS} = 4.82$$

5.6.2 Key Calculations

Consider a key with dimensions 8.00 mm wide (w), 7.00 mm deep (d) and 70.00 mm long.

The effective length (L_e) will be $70.00 - (2 \times 4.00 \text{ mm}) = 62.00 \text{ mm}$.

From Figure 5.3:

$$x = 2.94 \text{ mm} \qquad \text{By calculation.}$$

$$x_1 = 3.26 \text{ mm}$$

$$D_i \text{ (reduced shaft diameter)}$$

$$= D_o - t_1$$

$$= 25.00 - 4.00$$

$$= 21.00 \text{ mm}$$

The shear stress in the reduced shaft diameter:

$$\tau = \frac{16 \cdot T \cdot K_s}{\pi \cdot D_i^3}$$

$$= \frac{16 \times 63.66 \text{ Nm} \cdot 2}{\pi \times (21.00 \text{ mm})^3}$$

$$= 70.02 \text{ MPa}$$

$$\text{FoS} = \frac{\sigma_s}{\tau}$$

$$= \frac{200 \text{ MPa}}{70.02 \text{ MPa}}$$

$$= 2.86$$

5.6.3 STRAIGHT SPLINE CALCULATIONS

Consider a straight sided spline with an outside diameter of 25.00 mm.

The nomenclature used for a straight sided spline is as shown in Figure 5.4.

For this example see Table 5.7 for input data for straight sided spline calculations.

The shear stress in the reduced diameter (D_i):

$$\tau = \frac{16 \cdot T \cdot K_s}{\pi \cdot D_i^3}$$

$$= \frac{16 \times 63.66 \text{ Nm} \cdot 2}{\pi \times (21.00 \text{ mm})^3}$$

$$= 70.02 \text{ MPa}$$

FoS:

$$\text{FoS} = \frac{\sigma_s}{\tau}$$

$$= \frac{200 \text{ MPa}}{70.02 \text{ MPa}}$$

$$= 2.86$$

The compressive stress in the spline and shaft:

$$\sigma_c = \frac{T \cdot K_s}{L_e \cdot n \cdot r \cdot y}$$

$$= \frac{63.66 \text{ Nm} \times 2}{25.0 \text{ mm} \times 6 \times 11.5 \text{ mm} \times 1.4 \text{ mm}}$$

$$= 52.72 \text{ MPa}$$

TABLE 5.7

Input Data for Straight Sided Spline Calculation

D_i	= Reduced diameter	21.00	mm
n	= Number of teeth	6	
L_e	= Effective length	25.00	mm
r	= Mean radius of teeth	11.50	mm
c	= Chamfer/radius size at top and bottom of teeth	0.3 (assumed)	mm
y	= The effective depth of the teeth	1.40 ((D − Di)/2 − 2.c)	mm

FoS:

$$= \frac{130 \text{ MPa}}{52.72 \text{ MPa}}$$

FoS = 2.47

5.6.4 Involute Spline Calculations

Considering an involute spline of 25.00 × 24 teeth.

The nomenclature used for an involute spline is covered in Figure 5.5.

For this example see Table 5.8 for input data for involute sided spline calculations.

The shear stress in the reduced shaft diameter (D_i):

$$t = \frac{16 \cdot T \cdot K_s}{\pi \cdot D_i^3}$$

$$= \frac{16 \times 63.66 \text{ Nm} \times 2}{\pi \times (2250 \text{ mm})^3}$$

$$= 56.927 \text{ MPa}$$

FoS:

$$= \frac{200 \text{ MPa}}{56.927 \text{ MPa}}$$

FoS = 3.513

The shear stress in the spline teeth:

$$\tau = \frac{4 \cdot T \cdot K_s}{L \cdot n \cdot t \cdot D}$$

TABLE 5.8
Input Data for Involute Spline Calculation

n	= Number of teeth	24	mm
m	= Module	1.00	mm
L	= Length of spline	10.0	mm
D	= Mean (pitch) diameter	24.00 (m.n)	mm
Di	= Reduced diameter	22.50 (m.(n.1.5))	mm
Do	= Outside diameter	25.00 (m.(n + 1))	mm
p	= Pitch	3.1416 (m.π)	mm
t	= Tooth thickness	1.157 (p/2)	mm
h	= Tooth height	0.90 (0.9.m)	mm
c	= Chamfer/radius size at top and bottom of teeth	0.3 (assumed)	mm
y	= The effective depth of the teeth	1.40 ((D – Di)/2 – 2.c)	mm

$$= \frac{4 \times 63.66 \times 2}{10.00 \times 24 \times 57 \text{ mm} \times 24.00 \text{ mm}}$$

$$= 56.316 \text{ MPa}$$

FoS:

$$= \frac{200 \text{ MPa}}{56.316 \text{ MPa}}$$

$$\text{FoS} = 3.551$$

The compressive stress in the spline teeth:

$$\tau = \frac{2 \cdot T \cdot K_s}{L \cdot n \cdot h \cdot D_i}$$

$$= \frac{2.63 \times 66 \text{ Nm} \times 2}{10.00 \text{ mm} \times 24.0 \times 9 \text{ mm} \times 4.00 \text{ mm}}$$

$$= 49.12 \text{ MPa}$$

FoS:

$$= \frac{130 \text{ MPa}}{49.12 \text{ MPa}}$$

$$\text{FoS} = 2.647$$

Table 5.14 summarises the stresses calculated in the above examples.

Tables 5.9, 5.10, 5.11, 5.12 and 5.13 give the respective values for the factors K_d, K_a, K_m, K_f and K_w used in the calculations.

TABLE 5.9
Showing Keyway/Spline Design Factor K_d

Coupling Design	Design Factor
Fixed (close fit) loaded.	1
No relative axial/radial movement of the hub and shaft occurs: the relative positions are fixed using a suitable construction design, e.g. threaded fasteners, press fits.	
Flexible/sliding—unloaded.	3
Relative position of the shaft and hub is not fixed. Axial movements of the hub along the shaft occur only when coupling is unloaded.	
The shaft is effectively fixed when loaded.	
Flexible/sliding when loaded.	9
Sliding coupling with loading relative positions of the shaft and hub is not fixed. Axial shifts of the hub along shaft occur in loaded couplings.	

TABLE 5.10
Showing Spline Application Factors K_a

| | Type of Load | | | |
	Uniform Generators, Fans	Light Shock Oscillating Pumps	Intermediate Shock Actuators	Heavy Shock Presses, Shears
Power Source	Application Factor (K_a)			
Uniform (turbine motor)	1.0	1.2	1.5	1.8
Light shock (hydraulic motor)	1.2	1.3	1.8	2.1
Medium shock (ICE)	2.0	2.2	2.4	2.8

TABLE 5.11
Showing the Spline Distribution Factors K_m

| Misalignment (mm/mm) | Load Misalignment Factor (K_m) | | | |
	12 mm Face Width	25 mm Face Width	50 mm Face Width	100 mm Face Width
0.001	1.0	1.0	1.0	1.5
0.002	1.0	1.0	1.5	2.0
0.004	1.0	1.5	2.0	2.5
0.008	1.5	2.0	2.5	3.0

Note: This factor relates to the reduction in strength as a result of the misalignment of the spline.

TABLE 5.12
Showing the Fatigue Life Factors for Splines K_f

| Number of Torque Cycles (Start/Stop Cycles) | Fatigue Life Factor (K_f) | |
	Unidirectional	Fully Reversed
1×10^3	1.8	1.8
1×10^4	1.0	1.0
1×10^5	0.5	0.4
1×10^6	0.4	0.3
1×10^7	0.3	0.2

TABLE 5.13

Showing Wear Life Factor for Splines K_w

No. of Revolutions of Spline	Life Wear Factor (K_w)
10×10^3	4.0
10×10^4	2.8
10×10^5	2.0
10×10^6	1.4
10×10^7	1.0
10×10^8	0.7
10×10^9	0.5

TABLE 5.14

Summary of Key, Straight Sided and Involute Spline Strengths from Examples

	Key	Straight Sided Spline	Involute Spline
Reduced shaft shear stress	70.02 MPa	70.02 MPa	56.927 MPa
Factor of safety (FoS)	2.86	2.86	3.513
Compressive stress in spline teeth		52.72 MPa	49.12 MPa
Factor of safety (FoS)		2.47	2.647

6 Methods of Attachments

The attachment of brackets to structural members requires special attention depending on the nature of the attachment and the type of loading on the bracket.

Three types of attachment are considered here. These represent the most common form of fitting:

- Bolts in tension
- Bolts in shear
- Welding

6.1 BOLTS IN TENSION

Example 6.1

Consider a bracket depicted in Figure 6.1. The bracket is attached to the adjacent member using pairs of fasteners at positions a, b and c and is subject to vertical force P.

Solution:

$$\text{The direct loading on the bolt } = \frac{P}{n}$$

where
   P = total load
   n = number of bolts

When the line of action does not pass through the central area of the bolt configuration, the following procedure is used.

6.1.1 LOADING PRODUCING A TENSILE LOAD IN BOLT

The bracket is considered to "heel" about point o under the influence of the load at P.

By inspection it will be apparent that the row of bolts farthest away from the centre of rotation will be the most heavily loaded.

$$m = \frac{P \cdot L \cdot l_c}{\Sigma l^2} \tag{6.1}$$

$$P = 150 \text{ kN}$$

$$= 200 \text{ mm}$$

$$n = 6 \text{ (2 bolts per position)}$$

$$l_a = 20 \text{ mm}$$

$$l_b = 95 \text{ mm}$$

$$l_c = 170 \text{ mm}$$

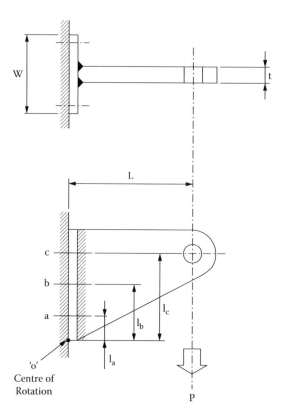

FIGURE 6.1   Detail of angle bracket: Example 6.1.

$$\Sigma l^2 = (l_a^2 + l_b^2 + l_c^2)$$

$$\Sigma l^2 = (20^2 + 95^2 + 170^2)$$

$$\Sigma l^2 = 38325 \text{ mm}^2$$

Maximum tension in bolts at position $l_c = F_{tension}$:

$$F_{tension} = \frac{150 \text{ kN} \times 200 \text{ mm}}{2 \times 225.44 \text{ mm}}$$

$$F_{tension} = 66.356 \text{ kN}$$

Total load on bolts $= F_{total}$

$$F_{total} = F_{tension} + \frac{P}{n}$$

$$F_{total} = 66.536 + \frac{150}{6}$$

$$= 91.536 \text{ kN}$$

Considering fastener diameter $= 30$ mm

$$\text{Core diameter} = 25.706 \text{ mm}$$

$$\text{Maximum stress} = \sigma_{max}$$

$$\sigma_{max} = \frac{91.536 \times 10^3}{\dfrac{\pi \times 25.706^2}{4}}$$

$$\sigma_{max} = 176.37 \text{ MPa}$$

6.1.1.1 Permissible Stress

Using fasteners to BS 3692 strength grade 4.6, yield strength = 235 N/mm². Therefore use 6 × 30 mm diameter bolts.

$$\text{FoS} = \frac{\sigma_{yield}}{\sigma_{max}}$$

$$= 1.332$$

Note: It is generally recognised that under high loads the heel point O will suffer a degree of plastic deformation and the true heel point will migrate towards the first row of fasteners at position a. Some argue that this is the point where the moment should be measured.

It is considered that the former position should be adopted, as the second position will have the effect of reducing the load on the fasteners.

6.1.2 LOAD PRODUCING A TENSION AND SHEAR LOAD IN BOLT

Example 6.2

Consider the bracket shown in Figure 6.2 where the applied force is in the plane of the connections. In this case the load produces a tension and shear force on the bolts.

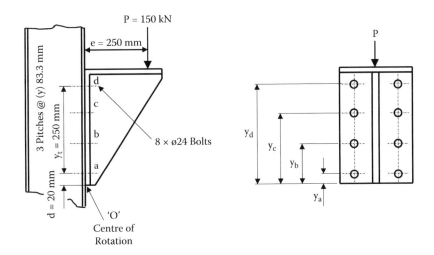

FIGURE 6.2 Angle bracket: Example 6.2.

Solution:

$P =$	150 kN	$y_a =$	20 mm + y
$e =$	250 mm	$y_b =$	20 mm + (2 × y)
	moment arm		
$y =$	83.3 mm pitch	$y_c =$	20 mm + (3 × y)
$Y =$	250 mm	$y_d =$	20 mm + (4 × y)
$n =$	8 bolts		
$d =$	20 mm		
$A =$	$\dfrac{\pi \cdot d^2}{4}$	$\Sigma y^2 =$	$(y_a^2 + y_b^2 + y_c^2 + y_d^2)$
$\therefore A =$	314.16 mm²	$\Sigma y^2 =$	243.087 × 10³ mm²

$$P_1 = P \times e$$

$$P_1 = 37.5 \times 10^3 \text{ kN/mm}^2$$

$$Y_1 = \frac{\Sigma y^2}{Y}$$

$$Y_1 = 900.321 \text{ mm}$$

The resultant stress on the top fastener is a vector quantity derived from the stress σ_t and the direct stress σ_s.

$$\sigma_r = (\sigma_{tension}^2 + \sigma_{shear}^2)^{0.5}$$

$$\sigma_r = (61.137^2 + 55.043^2)^{0.5}$$

$$\sigma_r = 82.265 \text{ MPa}$$

Tensile load:

$$F_t = \frac{P_1}{2 \times Y_1}$$

$$F_t = 20.826 \text{ kN}$$

Direct load:

$$F_s = \frac{P}{n}$$

$$F_s = 18.75 \text{ kN}$$

$$\sigma_{shear} = \frac{P}{n \times A}$$

$$\sigma_{shear} = \frac{150 \text{ kN}}{8 \times 340.645 \text{ mm}^2}$$

$$= 55.043 \text{ MPa}$$

$$\sigma_{tension} = \frac{F_t}{\dfrac{\pi \times 20.319^2}{4}}$$

$$\sigma_{tension} = \frac{20.826 \text{ kN}}{340.645 \text{ mm}^2}$$

$$= 61.137 \text{ MPa}$$

Maximum stresses:

$$\sigma_{yield} = 235 \text{ kN/mm}^2$$

Permissible stress: Using fasteners to BS 3692 strength grade 4.6, yield strength = 235 kN/mm². Factor of Safety (FoS):

$$FoS = \frac{\sigma_{allowable}}{F_r}$$

$$FoS = \frac{235 \text{ MPa}}{\sigma_r}$$

$$= 2.857$$

6.1.3 BOLTS IN SHEAR DUE TO ECCENTRIC LOADING

Example 6.3

When the line of action is in the plane of the joint as shown in Figure 6.3, the fasteners will be in shear.

If the line of action of the load passes outside the centre of area, the bracket will tend to turn and the fasteners will be eccentrically loaded.

Solution:

Let u_1 = load/fastener per unit distance from O due to the couple WL.

$$P \cdot L = 4u_1 l_a^2 + 2u_1 l_b^2$$

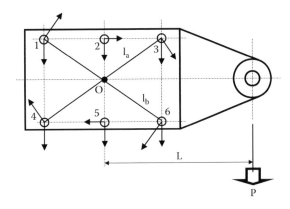

FIGURE 6.3 Bracket subject to eccentric loading.

Moments about O from which u_1 can be found:

$$P = 150 \text{ kN}$$

$$L = 225 \text{ mm}$$

$$L_a = 111.803 \text{ mm}$$

$$L_b = 111.803 \text{ mm}$$

$$n = 6$$

Re-arranging the above formulae to solve for u_1:

$$u_1 = \frac{P \cdot L}{\left[\left(4 \cdot L_a^2\right) + \left(2 \cdot L_b^2\right)\right]}$$

$$u_1 = 450.003 \text{ kN/m}$$

Load on fastener due to turning moment:

$$L_t = u_1 \cdot L_a$$

$$L_t = 50.312 \text{ kN}$$

Direct load on fastener due to load:

$$L_d = \frac{P}{n}$$

$$L_d = 25 \text{ kN}$$

Resultant load (solved using the sine and cosine rules):

$$c = \frac{L_t \times 0.894426}{0.6877428}$$

$$c = 65.432 \text{ kN (see Figure 6.4)}$$

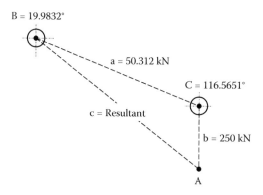

FIGURE 6.4   Resultant load for Example 6.3.

6.2 WELDING (PERMANENT)

6.2.1 STRENGTH OF WELDED JOINTS

Depending upon the angle between the fillets and line of action of the load, if reverse loads are experienced, these values must be modified and stress concentration factors can be used.

Table 6.1 gives the approximate strengths of some welded joints and Table 6.2 tabulates the stress concentration factors generated by different welds.

Example 6.4

Consider the welded bracket shown in Figure 6.5(a).

TABLE 6.1
Strength of Welded Joints

Type of Joint	Description of Stress Condition	Allowable Stress
Butt joints	Working stress for steel under tensile or compressive load	120–140 MPa
	Shear stress	80 MPa
Filler joints	Working stress–End fillets	
	Side fillets	95 MPa
	Diagonal fillets	80 MPa
		80–95 MPa

TABLE 6.2
Stress Concentration Factors of Various Welds

Type of Weld	Stress Concentration Factor
Reinforced butt	1.2
Toe of transverse fillet	1.5
T butt joint with sharp corners	2.0
End of parallel fillet	2.7

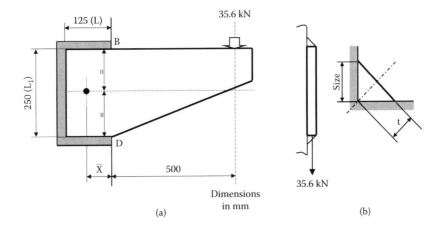

FIGURE 6.5 (a) Welded bracket for Example 6.4. (b) Weld throat dimensions.

The bracket is to be welded to a column and carries an eccentric load of 35.6 kN; find the size of weld required if the allowable weld stress is 77 MPa.

The weld is subjected to a direct shear force and an eccentric turning effect; note the 125 mm dimension is assumed to be the centroid of the vertical fillet weld.

Solution:

To find the position of the centroid of the weld, take moments about BD to find X, and let t = throat thickness of weld as shown in Figure 6.5(b) and Figure 6.6 and considering t = 1 (unity).

Table 6.3 gives the calculated sectional properties for the example weld.

$$\frac{\Sigma Ay}{\Sigma A} = \frac{46875}{500}$$

$$\overline{X} = 93.75 \text{ mm}$$

$$\overline{Y} = 125 \text{ mm}$$

The larger resultant stresses at positions B and D are the vectors of the vertical and horizontal components generated by the eccentricity of the load together with the direct shear, as shown in Figure 6.7.

The polar second moment of area (J) of the weld arrangement:

$$J = 2a\left(\frac{l^2}{12} + R^2\right) + a_1\left(\frac{l_1^2}{12} + R_1^2\right)$$

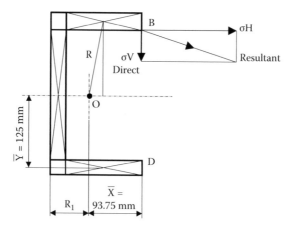

FIGURE 6.6 Centroid positions of welds in Example 6.4.

TABLE 6.3

Sectional Properties of Weld in Example 6.4

Section	b	d	A	y	Ay
1	125	1	125	62.5	7812.5
2	1	250	250	125	31250
3	125	1	125	62.5	7812.5
		$\Sigma A =$	500	$\Sigma Ay =$	46875

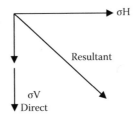

FIGURE 6.7 Resultant vector for Example 6.4.

$$= 2 \times 125t\left(\frac{125^2}{12} + 129^2\right) + 250t\left(\frac{250t}{12} + 31^2\right)$$

$$= 6.03 \times 10^6t \ mm^4$$

where
R = 129 mm
R_1 = 31 mm
Area A = 125 t
Area A_1 = 250 t
Stress at B and D:

$$\bar{x} = 93.75 \ mm$$

$$= \frac{Q}{A} + \frac{M}{J} \cdot y$$

$$= \frac{35.6 \ kN}{2 \times 125t + 250t} + \frac{35.6 \ kN \times 593.75 \ mm}{6.03 \times 10^6 t} \times 93.75 \ mm$$

$$= \frac{0.3998}{t}$$

Vertical σ = stress due to direct load + stress due to the eccentric load when

$$\sigma H = \frac{35.6 \ kN \times 593.75 \ mm \times 125 \ mm}{6.03 \times 10^6 t}$$

Horizontal σ when y = 125 mm:

$$= \frac{0.438 \ kN}{t \ mm^2}$$

The resultant:

$$\sigma = \sqrt{\left(\frac{0.3998}{t}\right)^2 + \left(\frac{0.438}{t}\right)^2}$$

$$= \frac{0.593 \ kN}{t \ mm^2}$$

The allowable working stress = 77 MPa = 0.077 kN/mm².

To find throat thickness t mm:

$$\frac{0.593}{t} = 0.077$$

$$t = \frac{0.593}{0.077}$$

$$t = 7.701 \text{ mm}$$

Size of weld = 1.414t

= 1.414 × 7.701 mm

= 10.89 mm

7 Columns and Struts

7.1 BACKGROUND

A column is vertical and supported at both ends. A strut may be inclined or even horizontal and have a variety of end fixings.

The notes below consider that the members are straight and manufactured from a homogeneous engineering material and are used within the elastic operating range of the material. It is further considered that the applied load or force is being applied along the centroid of the end features.

The column or strut will remain straight until the end force reaches a certain value and buckling begins. An increase in force will then result in the column further buckling, but a reduction in this force will then result in the column or strut returning to its original condition. The value of this critical force will depend upon the slenderness ratio and the end fixing conditions together with the material of construction.

The slenderness ratio (λ) is defined as:

$$\lambda = \frac{l}{k_{\min}} \tag{7.1}$$

where l is the effective length and $k_{\min}$ is the least radius of gyration of the section.

The principal end fixing conditions are as follows:

1. Pinned (hinged) at both ends
2. Fixed (built in) at both ends
3. Fixed at one end and free at the other end
4. Fixed at one end and pinned at the other end

Figure 7.1 depicts these end conditions.

The failure of a strut or column is a function of its length and will have a tendency to fail in pure bending; in this situation the Euler formula is suitable to analyse the condition.

$$\sigma_c = \frac{\pi^2 E}{\left(\dfrac{L}{k_{\min}}\right)^2} \tag{7.2}$$

Euler's theory takes no account of the compressive stress in the member. For a material with a compressive stress less than 300 MPa and a Young's modulus approximately 200 kPa, the strut will tend to fail in compression when the slenderness ratio (l/k) is less than 80. It has been found that Euler's equation is not reliable for slenderness ratios less than 80 and should be avoided with slenderness ratios less than 120.

In many practical cases struts may have slenderness ratios below which the Euler formula is applicable. A number of empirical formulae have been developed to improve the prediction of the critical stress, and these include, among others:

- Rankine-Gordon
- Perry-Robertson

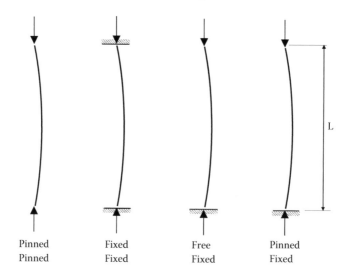

Pinned   Fixed   Free   Pinned
Pinned   Fixed   Fixed   Fixed

FIGURE 7.1   End fixing conditions.

- Johnson-Euler
- Euler-Engesser

The first two theories (Rankine-Gordon and Perry-Robertson) will be considered in this chapter.

7.2   RANKINE-GORDON METHOD

Gordon suggested an empirical formula be used (based on experimental data). Rankine modified this formula to the one used today.

Formula:

$$P_r = \frac{A\sigma_c}{1 + a\left(\dfrac{l}{k}\right)^2} \tag{7.3}$$

where

P_r = crushing or crippling load (Rankine-Gordon) (kN)

σ_c = direct crushing stress (MPa)

A = cross-sectional area of column or strut (m²)

l = length of strut or column (m)

k = least radius of gyration of cross section (m)

a = constant depending upon end fixing

also $k = \sqrt{\left(\dfrac{I}{A}\right)}$ and I the least second moment of area of section

Typical values for σ_c and a are shown in Table 7.1; these will vary dependent upon materials and type of end fixings.

The safe load:

$$P = \frac{P_r}{\text{Factor of safety}} \tag{7.4}$$

TABLE 7.1
Typical Values of a for Use in the Rankine-Gordon Formula

Material	σ_c (MPa)	Value of a		
		Fixed Ends	Hinged Ends	One End Fixed, the Other End Hinged
Cast iron	560	1/6400	1/1600	1/3600
Low carbon steel	325	1/30,000	1/7500	1/16,875

Note: Since the above values of a are not exactly equal to the theoretical values, the Rankine-Gordon loads for long columns or struts will not be identical to those estimated by the Euler theory as estimated.

For long columns, Euler's formula applies:

$$P_e = \frac{\pi^2 EI}{l^2} \text{ for hinged or rounded ends.}$$

where

$$P_e = \frac{\pi^2 EI}{4l^2} \text{ for one end fixed}$$

E = modulus of elasticity (GPa)

P_e = crippling load (Euler)

For short columns, where buckling effects are absent and hence material is in direct compression, these equations reduce to:

$$P_r = P_e = A\sigma_c \tag{7.5}$$

Example 7.1

A strut in a framed structure is manufactured from a steel pipe 150 mm outside diameter and 12.5 mm wall thickness. The length is 3.05 m and the pipe is pin-jointed at both ends.
    Using a Factor of Safety (FoS) of 5, what is the safe load?

Solution:

$$P_r = \frac{A\sigma_c}{1 + \left(\dfrac{l^2}{k^2} \right)}$$

$$A = \frac{\pi}{4}(0.15^2 - (2 \times 0.0125)^2)$$

$$= 0.00540 \text{ m}^2$$

$$\sigma_c = 325 \text{ MPa} \qquad \text{(from Table 7.1)}$$

$$a = \frac{1}{7500} \qquad \text{(for hinged ends)}$$

$$l = 3.05 \text{ m}$$

$$k^2 = \frac{I}{A} \qquad \text{(for a circular pipe)}$$

$$I = \frac{\pi}{64}(D^4 - d^4)$$

$$= \frac{\pi}{64}(0.15^4 - 0.125^4)$$

$$= \frac{\pi}{64}(0.000506 - 0.000244)$$

$$= 0.00001286 \text{ m}^4 \quad \text{or} \quad 12.861 \times 10^{-6} \text{ m}^4$$

and

$$FoS = 5$$

$$\text{Safeload} = \frac{1.154}{5}$$

$$= 0.2308 \text{ MN}$$

$$\text{say} = 231 \text{ kN}$$

Example 7.2

A column for a crane gantry consists of two 406 × 152 universal beams, connected by 12.5 mm thick plates as shown in Figure 7.2.

The overall length is 8.2 m and the ends may be considered fixed.

Calculate the safe load using a Safety Factor (SF) of 5.

Solution:

From the *Steel Designers' Manual*, for 406 mm × 140 mm × 46 kg universal beam the least second moment of area quoted:

$$I_{yy} = 539 \text{ cm}^4$$

$$A = 59 \text{ cm}^2$$

For the whole column, it can be shown that the least value of I is about the Y:Y axis.

$$I_{yy} = (I_{cg} + Ah^2) \times 2$$

$$= (539 + 59 \times 19^2) \times 2$$

$$= 43676 \text{ cm}^4$$

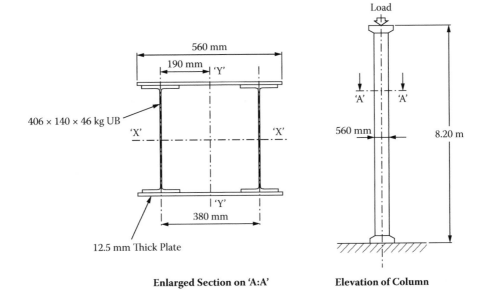

FIGURE 7.2 Graphic for Example 7.2.

To find I_{yy} for column, for two 406 mm × 140 mm sections:

$$I_{yy} \text{ for column} = 43676 \text{ cm}^4$$

$$\text{Area of cross section} = (2 \times 59) + (56 \times 2.5)$$

$$= 258 \text{ cm}^2$$

$$= 0.0258 \text{ m}^2$$

In this case the connecting plates can be disregarded and omitted from the second moment of area. Hence

$$k^2 = \frac{43676}{59 \times 2}$$

$$= 370.136 \text{ cm}^2$$

$$\text{and} \quad l = 8.2 \text{ m}$$

To evaluate k^2 the area of the connecting rods is ignored.
From Table 7.1, case I mild steel – fixed ends,

$$a = \frac{1}{30000}$$

and

$$\sigma_c = 325 \text{ MPa}$$

$$P_r = \cfrac{0.0258 \text{ m}^2 \times 325 \text{ MPa}}{1 + \cfrac{1}{30000} \times \cfrac{(8.2\,\text{m})^2}{0.03701\,\text{m}^2}}$$

$$= 7.906 \text{ MN}$$

FoS of 5:

$$\text{Safe load} = \frac{7.906 \text{ MN}}{5 \text{ (SF)}}$$

$$= 1.581 \text{ MN}$$

7.3   PERRY-ROBERTSON METHOD

The Perry-Robertson formula more closely represents the behaviour of practical columns allowing for the uncertainty of the eccentricity of loading, initial curvature pre-existing in the column, local defects, etc.

Here the critical value for stress is calculated from:

$$\sigma_c = \frac{\sigma_y + (\eta+1)\sigma_e}{2} - \sqrt{\left\{\left[\frac{\sigma_y + (\eta+1)\sigma_e}{2}\right]2 - \sigma_y.\sigma_e\right\}} \tag{7.6}$$

where

$$\eta = 0.003\frac{l}{k_{\min}}$$

Figure 7.3 compares the results of the Perry-Robertson calculations against the Euler's predictions for steel and aluminium.

Notes on the use of the Perry-Robertson method:

1. In practice $\dfrac{l}{k_{\min}} \le 85$ $\qquad\qquad$ (7.7)

2. When $\dfrac{l}{k_{\min}} = 85$ all steels give a crippling stress (σ_c) of 210 MPa (approximately).

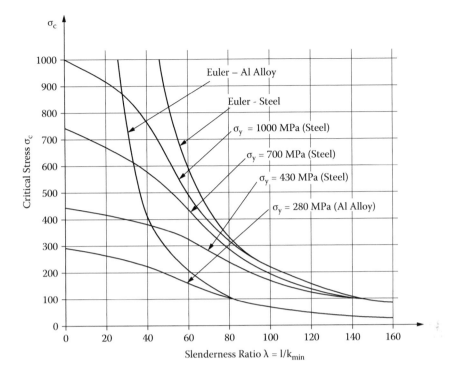

FIGURE 7.3 Comparison of Perry-Robertson against Euler's calculations.

8 Eccentric Loading

There are cases where the applied load acting on an element does not act on the neutral axis of a section. In this situation the principles of superposition will apply. Figure 8.1(a), (b), and (c) demonstrates the application. Figure 8.1(a) considers the element with an axial load acting directly on the neutral axis. Figure 8.1(b) then considers the effect of an eccentric load which is offset from the neutral axis; combining the loading of (a) and (b) will produce a load distribution as shown in (c).

It will be seen that the distribution of the load is offset where the loads are not equal about the neutral axis. The stresses in the element will reflect directly that of the load.

The general formulae for eccentric loading:

$$\sigma_1 = \frac{W_e}{A}\left(1 + \frac{X.Y_1}{k^2}\right)$$

$$\sigma_2 = \frac{W_e}{A}\left(1 - \frac{X.Y_2}{k^2}\right)$$

where

σ_1 = extreme fibre stress nearest load (MPa)

σ_2 = extreme fibre stress farthest from load (MPa)

W_e = actual eccentric load (kN)

A = area of section (m^2)

x = arm of eccentricity (m)

y_1 = perpendicular distance from neutral axis of section to outside edge nearest load (m)

y_2 = perpendicular distance from neutral axis of section to outside edge farthest from load (m)

k = radius of gyration (m)

The above may be demonstrated in the following example.

Example 8.1

Consider the cast iron shearing machine frame shown in Figure 8.2 which is subject to a force of 400 kN. A section of the frame is shown in Figure 8.3 on which is shown the position of the centroid, which has already been determined.

It is required to calculate the maximum tensile and compressive stresses in the section of the frame due to the given loading.

First find I_{xx} of the section; this will be the summation of parts (1), (2), (3) and (4). Use the parallel axis theorem for each part: This is determined as shown in Table 8.1. The next step is to determine the radius of gyration k^2.

Radius of gyration k^2:

$$k^2 = \frac{I}{A}$$

$$= \frac{897 \times 10^{-6} \ m^4}{52365 \times 10^{-6} \ m^2}$$

$$k^2 = 0.01713 \ m^2$$

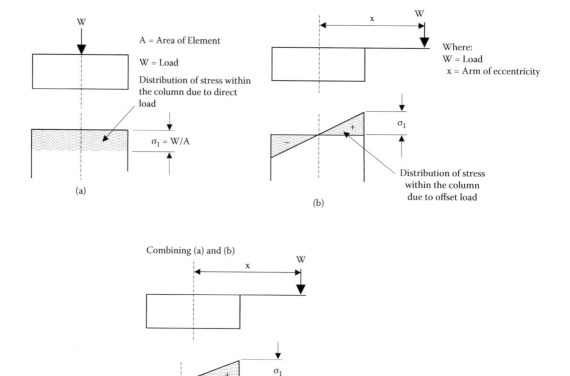

FIGURE 8.1 Principle of superposition.

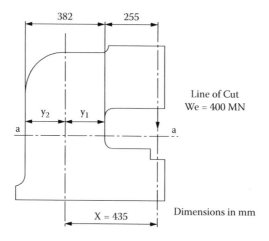

FIGURE 8.2 Outline elevation of frame.

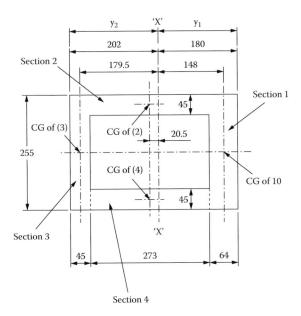

FIGURE 8.3 Enlarged section across section a:a.

The radius of gyration is left in this form as it is used in the subsequent calculations.

The load W_e will tend to open the gap in the frame. Therefore the maximum fibre stress is tensile on the edge nearest the load (section 1) and compressive on the edge farthest away from the load (section 3).

From Figure 8.1, $x = 0.435$ m.

From Table 8.1:

$$\text{Total moment of inertia } (I_{xx}) = 897 \times 10^{-6} \text{ m}^4$$

$$\text{Total area of the section} = 0.052365 \text{ m}^2$$

Maximum tensile stress:

$$\sigma_1 = \frac{W_e}{A}\left(1 + \frac{x \cdot y_1}{k^2}\right)$$

$$= \frac{400 \text{ kN}}{0.05237 \text{ m}^2}\left(1 + \frac{0.435 \text{ m} \times 0.180 \text{ m}}{0.01713 \text{ m}^2}\right)$$

$$\sigma_1 = 42.551 \text{ MPa}$$

Maximum compressive stress:

$$\sigma_1 = \frac{W_e}{A}\left(1 - \frac{x \cdot y_2}{k^2}\right)$$

$$= \frac{400 \text{ kN}}{0.05237 \text{ m}^2}\left(1 - \frac{0.435 \text{ m} \times 0.202 \text{ m}}{0.01713 \text{ m}^2}\right)$$

$$\sigma_2 = -31.541 \text{ MPa}$$

Note: If the direction of the loading was reversed, then obviously σ_1 would become compressive and σ_2 tensile.

TABLE 8.1
Moments of Inertia about Axis x:x for Example 8.1

Section	b	d	A	y	Ay	I_{xx} (b*d³/12)	h	h²	I_{xx} + Ah²
1	0.255	0.064	0.0163	0.032	0.0005	5.57056E-06	0.1508	0.0227	0.000377
2	0.045	0.273	0.0123	0.2005	0.0025	7.62991E-05	0.0177	0.0003	8.01E-05
3	0.255	0.045	0.0115	0.3595	0.0041	1.93641E-06	0.1767	0.0312	0.00036
4	0.045	0.273	0.0123	0.2005	0.0025	7.62991E-05	0.0177	0.0003	8.01E-05

S area = 0.0524 m²

$\bar{X}$ = 0.182828 m

S Ay 0.0096 m³

S I_{xx} = 0.000897 m⁴

Total I_{xx} = 897.0 × 10⁻⁶ m⁴

9 Flat Plates

The theory of plates covers from circular plates to flat plates under a number of edge constraint conditions from simply supported to rigid edge restraints.

Table 9.1 gives descriptions of various restraints and loading conditions for some common circular plates covering simply supported and fixed edges. Table 9.2 also considers some common loadings on flat rectangular plates, including simply supported and fixed edges.

For the purposes of this section, a rectangular plate rigidly held at the edges and subject to a pressure on one side will be studied, as this is considered to be a more common design problem.

Example 9.1

A rectangular opening 406 mm × 252 mm in a pressure vessel is subjected to an internal pressure of 6 bar. It is closed by means of a cast iron flat plate cover (see Figure 9.1), which has a joint 19 mm wide and is assumed to withstand the same pressure. Design the flange and cover assuming the maximum allowable stress for the plate is 27.6 MPa and the stress in the studs is 42 MPa (this allows for the stress concentration factor at the root of the thread). For more information regarding stress concentration factors, see *Peterson's Stress Concentration Factors*.

Solution:

The diameter and number of studs needs to be decided next. A provisional diameter of M16 will be assumed and this will give a core diameter of 13.546 mm.

$$\text{Area of the core} = 144.116 \times 10^{-6} \text{ m}^2$$

$$\text{Load acting on each stud} = 42 \text{ MPa} \times 144.116 \times 10^{-6} \text{ m}^2$$

$$= 6.053 \text{ kN}$$

The pressure is assumed to act at the centroid of the shaded triangular area and is resisted by the studs at the centroid of the pitch line. The distance S is the bending arm.

It was decided to use 14 studs to give a more symmetrical arrangement.

The dimensions of the flange will depend upon whether hexagonal or socket set screws are used.

In this example, hexagonal fasteners were chosen, as these would make the flange a little wider and increase the bending moment acting on the flange.

The cover thickness t is found by taking moments about the diagonal as shown in Figure 9.2.

$$\text{Number of studs} = \frac{77.256}{6.053}$$

$$= 12.763$$

$$\text{Total pressure on triangular area} = \frac{77.256 \text{ kN}}{2}$$

$$= 38.628 \text{ kN}$$

$$\text{Length of the diagonal 'B'} = \sqrt{480^2 + 315^2}$$

$$= 574.13 \text{ mm}$$

TABLE 9.1
Circular Flat Plates (Constant Thickness)

Case	Description	Graphic	Stress & Deflection Equations
1	Concentrate central load with edge simply supported		$\sigma = \dfrac{3W(1-v)}{2\pi t^2}\left(\dfrac{1}{v+1}+\log\dfrac{\sigma}{r_0}-\dfrac{1-v\cdot r_0^2}{v\cdot 4a^2}\right)$ $\delta\,\mathrm{max} = \dfrac{3Wa^2(1-v)(3+v)}{4\pi Et^3}$
2	Uniformly distributed load with edge simply supported		$\sigma = \dfrac{3w(3+v)}{8}\left(\dfrac{a^2}{t^2}\right)$ $\delta\,\mathrm{max} = \dfrac{3wa^4(1-v)(5+v)}{16\pi Et^3}$
3	Concentrated central load with outer edge fixed		For $a > r_0$ $\sigma_{max} = \dfrac{3W(3+v)}{2\pi t^2}\left(\log\dfrac{a}{r_0}+\dfrac{r_0^2}{4a^2}\right)$ $\delta\,\mathrm{max} = \dfrac{3Wa^2(1-v)}{4\pi Et^3}$
4	Uniformly distributed load with outer edge fixed,		$\sigma_{max} = \dfrac{3Wa^2}{4t^2}$ $\delta_{max} = \dfrac{3wa^4(1-v^2)}{16Et^3}$ $\quad$ $\delta_{max} = H\dfrac{3wa^4(1-n^2)}{16Et^3}$

For thicker flat circular plates having
$(t/a = 0.1,)$ multiply the deflection equation by
the constant $H = 1 + 5.72\,(t/a)^2$

Bending moment using bending arm S = 53.72 mm (see Figure 9.2)

$$38.626 \times 53.72 = 2074.99 \text{ kN}$$

$$BM = \sigma_z$$

$$2074.99 \text{ kN} = 27.6 \times 10^3 \text{ Z} \quad \text{where } Z = \frac{Bt^2}{6}$$

$$\frac{2074.99 \times 6}{0.57413 \times 27.6 \times 10^3} = t^2$$

$$0.0007854 \text{ m}^2 = t^2$$

$$t = 28.02 \text{ mm}$$

S = 53.72 mm by setting out to scale

Use a 30 mm thick plate or the nearest larger stock standard size.

TABLE 9.2
Rectangular Plates – Common Loading

Case	Description	Graphic	Stress & Deflection Equations
1	Rectangular plate uniformly loaded with outer edges simply supported	'S' indicates simply supported edges	$\sigma_{max} = \dfrac{\beta q b^2}{t^2}$ (at centre) $\delta_{max} = \dfrac{-\alpha q b^4}{Et^3}$ (at centre) q = load per unit area

a/b	1.0	1.2	1.6	2.0	3.0	4.0	5.0	∞
β	**0.2874**	**0.3762**	**0.5172**	**0.6102**	**0.7134**	**0.7410**	**0.7476**	**0.7500**
α	0.0444	0.0616	0.0906	0.1110	0.1335	0.1400	0.1417	0.1421

Case	Description	Graphic	Stress & Deflection Equations
2	Uniform over a small concentric circle of radius r_o, outer edges simply supported		$\sigma_{max} = \dfrac{3W}{2\pi t^2}\left[(1+\nu)\ln\dfrac{2b}{\pi r'_o} + \beta\right]$ Note: $r'_o = \sqrt{1.6r_o^2 + t^2} - 0.675t$ if $r_o < t$ and $r'_o = r_o$ if $r_o > 0.5t$ $\delta_{max} = \dfrac{\alpha W b^2}{Et^3}$

Case	Description	Graphic	Stress & Deflection Equations
3	Uniformly distributed load with outer edges fixed		$\sigma_{max} = \dfrac{\beta q b^2}{t^2}$ (at centre of plate) q = load per unit area $\delta_{max} = \dfrac{\alpha q b^2}{Et^2}$ (at centre of plate)

a/b	1.0	1.2	1.4	1.6	1.8	2.0	∞
β	**0.1386**	**0.1794**	**0.2094**	**0.2286**	**0.2406**	**0.2472**	**0.2500**
α	0.0138	0.0188	0.0226	0.0251	0.0267	0.0277	0.0284

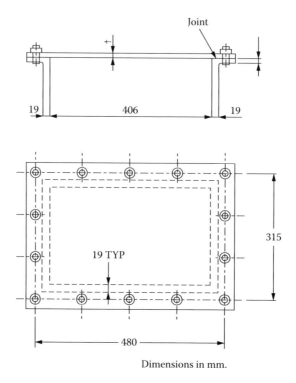

Dimensions in mm.

FIGURE 9.1   Flat plate cover for Example 9.1.

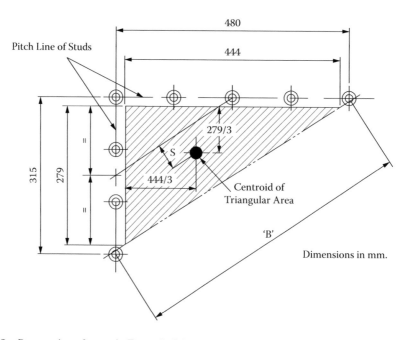

FIGURE 9.2   Part section of cover in Example 9.1.

To find the flange thickness the method used here is to take the load acting on one stud producing bending on a section of flange of width of one pitch.

$$\text{Minimum pitch} = \frac{315 \text{ mm}}{3}$$

$$= 105 \text{ mm}$$

$$\text{Load on section} = \frac{77.3}{14} \text{ kN}$$

$$= 5.52 \text{ kN}$$

$$\text{Bending arm} = 18 \text{ mm}$$

$$BM = \sigma_2$$

$$5.52 \text{ kN} \times 0.018 \times 10^3 \, m = \frac{\sigma b d^2}{6}$$

$$99.36 \text{ Nm} = 27.6 \times 10^6 \times \frac{0.105}{6} \, d^2$$

$$d = 0.0143 \text{ m}$$

To obtain sufficient depth of thread for the studs in the flange this should be 1.5 D = 24 mm.

Example 9.2

A 200 mm steel tube 9.5 mm thick has the ends closed by a flat steel end cover which is held in place by a central bolt and hexagonal nut, as shown in Figure 9.3. The pressure in the tube is 14 bar and the pressure on the joint is twice this value.

Find the bolt diameter and the cover plate thickness allowing for a working stress of 70 MPa. The bolt diameter may be neglected in this example.

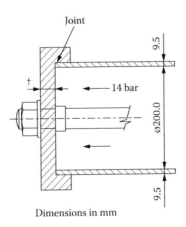

Dimensions in mm

FIGURE 9.3 End plate for Example 9.2.

Solution:

$$\text{Total bolt load for semi-circle} = \frac{61.5 \text{ kN}}{2}$$

$$= 30.75 \text{ kN}$$

$$\text{Load on } \varnothing 200 \text{ mm semi-circle area} = \frac{44 \text{ kN}}{2}$$

$$= 22 \text{ kN}$$

Load due to joint pressure $= 2 \times 14 \text{ bar} \times 10^5 \text{ N/m}^2 \times (\pi \times 0.2095 \text{ m} \times 0.0095 \text{ m})$

where 0.2095 m is the mean diameter of the tube.

$$\text{Load on end cover} = 14 \text{ bar} \times 10^5 \text{ N/m}^2 \times \frac{\pi}{4} \times 0.20^2 \text{ m (N)}$$

$$= 43.982 \text{ kN}$$

and 0.0095 m is the wall thickness.

$$= 17.507 \text{ kN}$$

Hence

$$\text{Total load on bolt} = 43.982 \text{ kN} + 17.507 \text{ kN}$$

$$= 61.489 \text{ kN}$$

Using the allowable stress of 70 MP:

$$= \frac{61.489 \text{ kN}}{70 \text{ MPa}}$$

$$= 0.0008784 \text{ m}^2$$

Area at root of thread of bolt:

$$\text{Area} = \frac{\pi d^2}{4}$$

$$0.000878 \text{ m}^2 = 0.7854 \, d^2$$

$$0.00119 = d^2$$

$$d = 0.03344 \text{ m}$$

$$= 33.44 \text{ mm root diameter}$$

Taking moments about the effective diameter of the end cover, the total load tending to bend the cover is the sum of half the load on Ø200 mm opening plus half the joint load, and acts at the centroids in each case.

These give a single resultant force, which is opposed by the central bolt, thus producing a couple balanced by the stress couple across the section of the cover.

$$\text{Total bolt load for semi-circle} = \frac{61.5 \text{ kN}}{2}$$

$$= 30.75 \text{ kN}$$

$$\text{Load on Ø200 mm semi-circle area} = \frac{44 \text{ kN}}{2}$$

$$= 22 \text{ kN}$$

$$\text{Load on joint for semi-circle} = \frac{17.5 \text{ kN}}{2}$$

$$= 8.75 \text{ kN}$$

Figure 9.4 illustrates the loading configurations:

For the semi-circle area centroid = 0.4244 × radius.
For the semi-circle of joint = 0.636 × radius.

$$\text{Centroid of Ø200 mm semi-circle area:}$$

$$= 0.4244 \times 100 \text{ mm}$$

$$= 42.4 \text{ mm}$$

$$\text{Centroid of joint area} = 0.636 \times 104.75 \text{ mm}$$

$$= 66.621 \text{ mm}$$

To find the position of the resultant, take moments about axis X:X (Figure 9.4).

$$8.75 \text{ kN} \times 66.621 \text{ mm} + 22 \text{ kN} \times 42.4 \text{ mm}$$

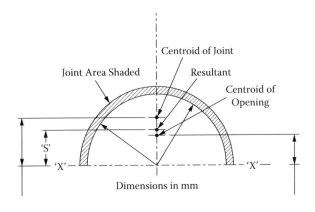

FIGURE 9.4 Loading arrangement for Example 9.2.

$$= 30.75 \text{ kN} \times S$$

$$1515.733 \text{ kNmm} = 30.75 \times S$$

$$49.292 \text{ mm} = S$$

Bending moment at X:X $= 30.75 \text{ kN} \times 0.0493 \text{ m}$

$$= 1.516 \text{ kNm}$$

Net width at X:X $= 219 \text{ mm} - 39 \text{ mm}$ (bolt diameter)

$$= 180 \text{ mm}$$

$$= 0.180 \text{ m}$$

$$1.516 \text{ kNm} = 70 \text{ MPa} \times \frac{0.180 \text{ m} \times t^2}{6} \quad \text{i.e.} \left(\frac{bd^3}{12} \right)$$

$$\frac{1.516 \times 6}{0.180 \times 70 \times 10^3} = t^2$$

$$t^2 = 0.0007219 \text{ m}^2$$

$$t = 0.02686 \text{ m}$$

$$t = 26.86 \text{ mm}$$

Say, $t = 27$ mm

10 Thick Cylinders

When analysing thin cylinders, the assumption made is that the stress in the wall of the cylinder will be uniform across the thickness when the cylinder is under internal pressure, and that the radial stresses will be negligible in comparison with the circumferential (hoop) and longitudinal stresses.

When the thickness of the cylinder is appreciable in relation to the diameter, this assumption cannot be justified and the variation in radial and circumferential stresses across the thickness can be deduced using Lamé's theory.

$$t = \frac{d}{2}\left\{\sqrt{\frac{\sigma + P}{\sigma - P}} - 1\right\} \tag{10.1}$$

where
  t = thickness (m)
  d = internal diameter (m)
  σ = maximum hoop stress (Pa)
  P = radial pressure (Pa)

Lamés theory
General formulae for shrink or force fits:

$$\frac{\delta}{2} = \frac{R}{E}\left(\sigma_{A_1} - \sigma_{B_2}\right) \tag{10.2}$$

For a cylinder subject to internal pressure (see Figure 10.1):

$$\sigma_{A_1} = P\left(\frac{r_o^2 + r_i^2}{r_o^2 - r_i^2}\right) \quad \text{Maximum stress} \tag{10.3}$$

where

$$\sigma_{B_1} = \frac{2Pr_i^2}{r_o^2 - r_i^2} \tag{10.4}$$

For a cylinder subject to external pressure (see Figure 10.2):

$$\sigma_{A_2} = \frac{2Pr_2^2}{r_o^2 - r_i^2} \quad \text{(Maximum stress)} \tag{10.5}$$

$$\sigma_{B_2} = -P\left(\frac{r_o^2 + r_i^2}{r_o^2 - r_i^2}\right) \tag{10.6}$$

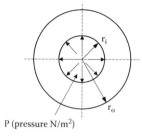

P (pressure N/m²)

FIGURE 10.1   Thick cylinder subject to internal pressure.

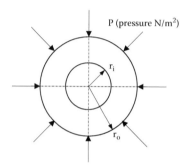

FIGURE 10.2   Thick cylinder subject to external pressure.

where
r_i = internal radius (m)
r_o = external radius (m)
t = thickness $(r_o - r_i)$ (t).
σ = maximum hoop stress N/m²
P = pressure (N/m²)

When two components are of different materials the above formula can be written as:

$$\frac{\delta}{2} = \frac{R}{E}\left(\sigma_{A_1} - v_1 P\right) - \frac{R}{E}\left(\sigma_{B_2} - v_2 P\right) \tag{10.7}$$

E_1 and E_2 = modulus of elasticity for the two different materials
v_1 and v_2 = Poisson's ratio for the two different materials

Example 10.1

A cast steel cylinder for a hydraulic press has an inside diameter of 230 mm. The internal pressure is 300 bar. Using a maximum hoop stress of 85 MPa, determine the thickness of the cylinder.

Solution:

From Lamé's formula:

$$t = \frac{d}{2}\left\{\sqrt{\frac{\sigma + P}{\sigma - P}} - 1\right\} \tag{10.8}$$

$$t = \frac{0.230 \text{ m}}{2} \left\{ \sqrt{\left(\frac{85 \times 10^6 \text{ N/m}^2 + 30.0 \times 10^6 \text{ N/m}^2}{85 \times 10^6 \text{ N/m}^2 - 30 \times 10^6 \text{ N/m}^2} \right)} - 1 \right\}$$

$$t = 0.05129 \text{ m}$$

Example 10.2

A cast iron hydraulic cylinder has an internal diameter of 150 mm and an external diameter of 200 mm. Internal pressure is 52 bar. Calculate the maximum hoop stress. Also determine the stress at the outer surface.

Solution:

The maximum hoop stress occurs at the internal surface:

$$\sigma_{A_1} = P \left(\frac{r_2^2 + r_1^2}{r_2^2 - r_2^2} \right)$$

$$= 52 \times 10^5 \text{ Pa} \times \left(\frac{0.1 \text{m}^2 + 0.075 \text{m}^2}{0.1 \text{m}^2 - 0.75 \text{m}^2} \right)$$

$$= 18.571 \text{ MPa}$$

Now stress at the outer surface:

$$\sigma_{A_1} = \left(\frac{2 P r_1^2}{r_2^2 - r_1^2} \right)$$

$$= \left(\frac{2 \times 52 \times 10^5 \times 0.075^2}{0.1^2 - 0.075^2} \right)$$

$$= 13.371 \text{ MPa}$$

Example 10.3

A mild steel universal coupling is shrunk on to a mild steel shaft Ø200 mm (see Figure 10.3). The shrink allowance is 1 mm per metre on diameter.

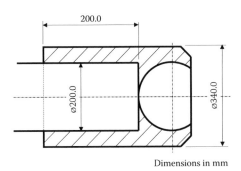

Dimensions in mm

FIGURE 10.3 Universal coupling for Example 10.3.

Calculate:

    a. The stresses in the shaft and coupling.
    b. For a maximum torque of 63 kNm, find the Factor of Safety (FoS) between the actual torque and the theoretical torque capable of being transmitted.

Solution:

    a. Coupling end
    where:
        $r_i = 100$ mm
        $r_o = 170$ mm
        at inner radius:

$$\sigma_{A_1} = P\left(\frac{r_o^2 + r_i^2}{r_o^2 - r_i^2}\right)$$

$$= P\left(\frac{0.170^2 + 0.100^2}{0.170^2 - 0.100^2}\right)$$

$$= 2.058\ P$$

        at outer radius:

$$\sigma_{B_1} = \frac{2Pr_i^2}{r_o^2 - r_i^2}$$

$$= \frac{2 \times P \times 0.100^2}{0.170^2 - 0.170^2}$$

$$= 1.058\ P$$

Consider the shaft:
For a solid shaft the hoop stress at the outside radius is equal to the radial stress since $r_i = 0$; then $\sigma_{B_2} = -P$.

    σ_{B_2} is used to correspond to the symbol in the formula for a hollow shaft or cylinder externally loaded.

$$\frac{\delta}{2} = \frac{r}{E}\left(\sigma_{A_1} - \sigma_{B_2}\right)$$

where
    δ = total shrink allowance 1 mm/m $\times$ 0.2 m
      = 0.2 mm
    r = 100 mm
    $E = 200 \times 10^9$ N/m² (for steel)

Substitute σ_{A_1} and σ_{B_2} in terms of P and write the units in terms of mm.

$$\text{Hence} \quad \frac{0.2\ \text{mm}}{2} = \frac{100\ \text{mm}}{200 \times 10^3\ \text{N/mm}^2}\left(2.058\ P - (-1.058\ P)\right)$$

$$0.1\ \text{mm} = \frac{100\ \text{mm}^3}{200 \times 10^3\ \text{N/mm}^2} \times \left(2.0058\ P + 1.058\ P\right)$$

$$0.1\ \text{mm} = 0.0005\ \text{mm}^3/\text{N} \times (3.116\ P)$$

$$\frac{0.10 \text{ N}}{0.005 \text{ mm}^2} = 3.116 \text{ P}$$

$$200 \text{ N/mm}^2 = 3.116 \text{ P}$$

$$\frac{200 \text{ N/mm}^2}{3.116} = P$$

$$64.185 \text{ N/mm}^2 = P \text{ or } 64.185 \text{ MPa (Radial Stress)} \qquad 1 \text{ N/mm}^2 = 1 \text{ MPa}$$

$$\sigma_{A_1} = 2.058 \times P$$

$$= 132.093 \text{ MPa}$$

$$\sigma_{B_1} = 1.058 \times P$$

$$= 67.908 \text{ MPa}$$

b. Safety Factor (SF)

These are the maximum and minimum hoop stresses generated in the coupling end due to the shrink fit on the solid shaft.

For the shaft $\sigma_{B2} = P = 64.185$ MPa.

The permissible maximum hoop stress should be within the elastic limit for the material between 432 and 494 MPa.

For this example the FoS will be:

$$\text{FoS} = \frac{\text{allowable stress}}{\text{working stress}}$$

$$\text{FoS} = \frac{432 \text{ MPa}}{132.093 \text{ MPa}}$$

$$= 3.270$$

The radial stress P is the gripping force between the coupling end and the shaft, if the radial stress P = 64.185 MPa and assuming a co-efficient of friction of 0.25 between the two surfaces:

The radial stress acts on a surface area:

$$= \pi DL$$

$$= \pi \times 0.2 \text{ m} \times 0.2 \text{ m}$$

$$= 0.1257 \text{ m}^2$$

Therefore the tangential force:

$$= 64.185 \text{ MPa} \times 0.126 \text{ m}^2 \times 0.25$$

$$= 2.0164 \text{ MN}$$

The shaft radius (r_i):

$$= 0.1 \text{ m}$$

Therefore the slippage torque will be:

$$= 2.0164 \text{ MN} \times 0.1 \text{ m}$$

$$= 201.64 \text{ kNm}$$

This is the theoretical maximum torque capable of being transmitted without slippage due to the gripping force of the shrinkage.

$$\text{Actual torque} = 63 \text{ kNm}$$

The FoS will be:

$$= \frac{201.64 \text{ kNm}}{63 \text{ kNm}}$$

$$= 3.20 \text{ (Ans.)}$$

11 Energy Formulae

11.1 FLYWHEELS BASICS

Flywheels are a convenient way to store mechanical energy. They are used in internal combustion engines to even out the fluctuating power strokes of the engine and in shearing presses to give an energy intensification during a cutting stroke and are found in a number of high tech energy conservation systems used on passenger road vehicles as a means of storing energy recovered from braking. Flywheels are designed to have the majority of their mass concentrated in the rim to store the maximum kinetic energy. The downside is the stresses in the rim of the flywheel will be increased, and if the flywheel exceeds its critical safe speed, then a catastrophic failure will result. Consider the simple flywheel depicted in Figure 11.1.

Basic energy formulae:

$$\text{Rotational kinetic energy} = \tfrac{1}{2}I\omega^2$$

$$\text{Moment of inertia} = mk^2$$

$$\text{Change in energy} = \tfrac{1}{2}I(\omega_1^2 - \omega_2^2)$$

$$\text{Torque} = I\alpha$$

$$\text{Power} = T\omega$$

where

I = moment of inertia (kgm²) about axis of rotation
m = mass of flywheel (kg)
k = radius of gyration (m) about axis of rotation
T = torque applied (Nm)
α = angular acceleration (rad/s²)

Table 11.1 gives some basic equations used in the design of flywheel systems.

To determine the size of a flywheel (I_z) necessary to ensure the speed is maintained within a specified range (i.e. the angular velocity **is** limited to a range ω_{min} to ω_{max} when the kinetic energy level varies between E_{min} and E_{max}):

Coefficient of speed fluctuation:

$$C_s = 2\frac{(\omega_{max} - \omega_{min})}{(\omega_{max} + \omega_{min})} = \frac{(\omega_{max} - \omega_{min})}{\omega_{mean}}$$

where

$$w_{mean} = \frac{\omega_{max} + \omega_{min}}{2}$$

$$E_{max} - E_{min} = \frac{I_z\left(\omega_{max}^2 - \omega_{min}^2\right)}{2}$$

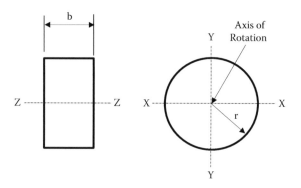

FIGURE 11.1 Simple flywheel.

TABLE 11.1

Basic Equations for Flywheel Systems

Case	Description	Symbol	Equation	Units
1	Angular velocity	(ω)	$=\dfrac{2\pi n}{60}$	$\left(\dfrac{rad}{sec}\right)$
2	Angular acceleration	(α)	$=\dfrac{(\omega_1 - \omega_2)}{t}$	$\left(\dfrac{rad}{s^2}\right)$
3	Angular acceleration	(α)	$=\dfrac{T}{I_z}$	$\left(\dfrac{rad}{s^2}\right)$
4	Energy stored	(E)	$=\dfrac{I_z . w^2}{2}$	N·m (Joules)
5	Change in kinetic energy when angular velocity changes from ω_1 to ω_2		$E_2 - E_1$ $=\dfrac{I_z\left(w_2^2 - w_1^2\right)}{2}$	N·m (Joules)

Assuming a flywheel is accelerated by application of a driving torque T_1 between angles θ_1 and θ_2 and decelerated by resisting its motion with a torque T_0 between angles θ_3 and θ_4:

6	The energy input to the flywheel (work input) between angles θ_1 and θ_2	$U_1 = T_1\,(\theta_2 - \theta_1)$	
7	The energy from a flywheel (work output) between angles θ_1 and θ_2	$U_0 = T_0\,(\theta_4 - \theta_3)$	

$$= I_z \frac{\left(\omega_{max} - \omega_{min}\right) - \left(\omega_{max} + \omega_{min}\right)}{2}$$

$$= C_s . I_z . w_{mean}^2$$

Hence

$$I_z = \frac{E_{max} - E_{min}}{C_s . \omega_{mean}^2}$$

Stresses generated within flywheel section due to speed of rotation: A flywheel section can be simplified to that of a rotating ring, enabling the determination of the stresses due to the rotation.

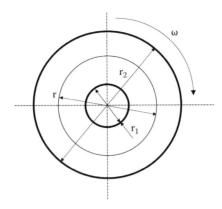

FIGURE 11.2 Tangential and radial stresses in a flywheel.

The theory involving the stresses in a thick walled cylinder may be used; the primary difference will be due to the inertial forces acting on the ring section.

The following conditions will apply in using the above assumption:

- The outside radius of the ring is greater than its thickness
- The section is constant
- The stresses are constant across its width

The tangential and radial stresses at radius r resulting from the rotation of a ring (see Figure 11.2) with an outside radius of r_2 and an inside radius of r_1 and rotating at an angular velocity (ω) will be found using the following equations.

Tangential tensile inertial stress:

$$\sigma_t = \frac{\rho \omega^2}{8} \left[(3+v) \left(r_1^2 + r_2^2 + \frac{r_1^2 \cdot r_2^2}{r^2} \right) - (1+3v) r^2 \right]$$

Radial tensile inertial stress:

$$\sigma_r = \rho \omega^2 \left(\frac{(3+v)}{8} \right) \left(r_1^2 + r_2^2 - \frac{r_1^2 \cdot r_2^2}{r^2} - r^2 \right)$$

The maximum tangential stress occurs in the bore and is equal to:

$$\text{Max } \sigma_t = \frac{\rho \omega^2}{4} \left[(1-v) r_1^2 + (3+v) r_2^2 \right]$$

The maximum radial stress will occur at a radius $r = \sqrt{(r_1 \cdot r_o)}$ and is equal to:

$$\text{Max } \sigma_r = \frac{\rho \omega^2}{8} (3+v)(r_2 - r_1)^2$$

Example 11.1

A single cylinder engine working on the four stroke cycle develops 9 kW at 5.8 rev/s. Find a suitable size of a cast iron flywheel if the energy absorbed is to be 0.8 of the energy developed per cycle.

Take the speed fluctuation as 0.5% above and below the mean. The average rim stress should not exceed 7M Pa.

Solution:

$$\text{Density of cast iron} = 7200 \text{ kg/m}^3$$

$$\text{Work done per rev} = \frac{\text{power}}{\text{working cycles/s}}$$

$$\text{4 stroke working cycles per second} = \frac{5.8 \text{ rev/s}}{2}$$

$$\text{Work done per working cycle} = \frac{9000 \text{ W}}{5.8/2 \text{ per sec}}$$

$$= 3103.44 \text{ Nm}$$

$$\text{Energy to be stored in flywheel} = 0.8 \times 3103.44 \text{ Nm}$$

$$= 2482.76 \text{ Nm}$$

To find the permissible rim velocity to generate a stress of 7 MPa,

$$\sigma = \rho\omega^2 r^2$$

where
  σ = centrifugal stress (N/m²)
  ρ = density of material (kg/m³)
  ω = angular velocity (rad/s)
  r = radius of wheel (m)

$$7 \times 10^6 \, \frac{N}{m^2} = 7200 \frac{kg}{m^3} \times 36.442 \frac{rad^2}{s^2} \times r^2 m^2$$

$$\frac{7 \times 10^6}{7200 \times 36.442^2} = r^2 \quad \left(\text{where } \frac{kg \times m}{s^2} \right) = N$$

$$0.7321 = r^2 \quad \text{and} \quad \omega = \text{rev/s} \times 2\pi$$

$$r = 0.8556 \text{ m (855.6 mm)}$$

$$\text{Diameter of flywheel} = 1.71 \text{ m}$$

$$\text{Change in energy} = \frac{I}{2}\left(\omega_1^2 - \omega_2^2\right)$$

$$2482.76 \text{ Nm} = \frac{I}{2} \times \left(36.6242^2 - 36.2598^2\right)$$

So,

$$0.5\% \text{ up on } 36.442 \text{ rad/s} = 36.6242 \text{ rad/s}$$

$$0.5\% \text{ down on } 36.442 \text{ rad/s} = 36.2598 \text{ rad/s}$$

$$2482.76 \, \text{Nm} = \frac{1}{2}(36.6242 + 36.2598) \times (36.6242 - 36.2598)$$

$$2482.76 \, \text{Nm} = \frac{1}{2}(26.559)$$

$$\frac{2 \times 2482.76}{26.559} = I$$

$$186.962 \, \text{Nms}^2 = I$$

$$\text{or } 186.962 \, \text{kgm}^2 = I$$

$$\text{now} \quad I = Mk^2$$

where:

$$k = \frac{D}{2} \text{ for a solid disc.}$$

Hence

$$k = \frac{1.71}{2}$$

$$k = 0.855 \, \text{m}$$

$$186.962 \, \text{kgm}^2 = M \times 0.855^2 \, \text{m}^2$$

Therefore $\quad M(\text{mass}) = 255.75 \, \text{kg}$

now $\quad \text{Volume} = \dfrac{\text{mass}}{\text{density}}$

$$= \frac{255.75 \, \text{kg}}{7200 \, \text{kg/m}^3}$$

$$= 0.0355 \, \text{m}^3$$

$$\text{Mean circumference} = \pi D$$

$$= 1.71\pi$$

$$= 5.372 \, \text{m}$$

$$\text{Cross-sectional area of rim} = \frac{0.0355 \, \text{m}^3}{5.372 \, \text{m}}$$

$$= 0.0066 \, \text{m}^2$$

$$\text{Area} = b \times t \, (\text{breadth} \times \text{thickness})$$

Some proportion should be considered between the breadth and thickness, say:

$$b = 1.2t$$

$$0.0066 \text{ m}^2 = 1.2 \text{ t}^2$$

$$0.0055 = t^2$$

$$t = 0.0742 \text{ m}$$

$$t = 74.16 \text{ mm}$$

and

$$b = 1.2t$$

$$b = 89.0 \text{ mm}$$

Example 11.2

A motor generator set for a rolling mill, running at 8.5 rev/s has a cast steel flywheel designed to take up the fluctuations of demand. The rotating parts of the motor and generator have a combined mass of 20 tonnes and a radius of gyration of 460 mm, and the flywheel has a mass of 50 tonnes and a radius of gyration of 1.6 m. If an excess of energy of 11 MNm is required when the mill is running over that supplied by the motor, calculate the decrease in speed.

Solution:

Total moment of inertia:

$$I = mk^2 \text{ for motor and generator} + mk^2 \text{ for flywheel}$$

$$= 20 \times 10^3 \text{ kg} \times 0.460^2 \text{ m}^2 + 50 \times 10^3 \text{ kg} \times 1.6^2 \text{ m}^2$$

$$= 4232 \text{ kgm}^2 + 128000 \text{ kgm}^2$$

$$= 132232 \text{ kgm}^2$$

$$\text{Change in energy} = \frac{I}{2}\omega_1^2 - \frac{I}{2}\omega_2^2$$

If R = the new speed (rev/s), then:

$$11 \times 10^6 \text{ Nm} = \frac{1}{2} \times 132232 \times \left\{ 8.5^2 \times (2\pi)^2 - R^2 (2\pi)^2 \right\}$$

$$11 \times 10^6 \text{ Nm} = \frac{1}{2} \times 132232 \times 4\pi^2 \left(8.5^2 - R^2 \right)$$

$$4.213 = 8.5^2 - R^2$$

$$R^2 = 72.25 - 4.213$$

$$R = 8.248 \text{ rev/s}$$

Decrease in speed = 8.5 − 8.248 rev/s

= 0.252 rev/s

Example 11.3

A flywheel has a mass of 2.03 tonnes and a radius of gyration of 1.22 m. It is keyed to the crank-shaft of an engine which develops 15 kW running at 2.5 rev/s. Assuming the mean torque constant at all speeds and neglecting all masses but the flywheel, find the time required to get the engine to full speed.

Solution:

Average torque:

$$T = \frac{power}{2\pi N}$$

$$= \frac{15 \times 10^3}{2\pi \times 2.5} \frac{Nm}{s} \times \frac{s}{rev}$$

$$= 954.93 \ Nm$$

$T = I\alpha$ where $\alpha = \omega/t$ and

$$I = mk^2$$

$$= 2.03 \times 10^3 \times 1.22^2 \ kgm^2$$

$$= 3021.45 \ kgm^3$$

$$T = I\alpha$$

$$954.93 \ Nm = 3021.45 \ kgm^2 \times \frac{2.5 \times 2\pi}{t} \frac{kgm^2}{s}$$

$$t = 49.72 \ s$$

Example 11.4

A cast iron flywheel for a shearing machine has a rim section of 240 mm wide. The outside diameter is 1.7 m and the inside diameter is 1.22 m. The flywheel has a vee-belt drive from an electric motor running at 10 rev/s. The motor pulley is 480 mm diameter. The energy absorbed during the effective cutting stroke of 75 mm is 80% of the energy supplied. When cutting a section, the speed of the flywheel is reduced by 0.25 rev/s. Determine the average cutting force exerted at the blade. Taking the shearing strength of the steel as 350 MPa, find the area of the steel being sheared. The inertia of the flywheel spokes and the rotating parts of the motor may be neglected for the purposes of this example. The density of cast iron is 7200 kg/m^3.

Solution:

$$Mass \ of \ rim \ = \ \frac{\pi}{4}(1.7^2 - 1.22^2) \times 0.24 \times 7200 \ m^2 \times m \times \frac{kg}{m^3}$$

$$= 1902.21 \ kg$$

$$\frac{\text{Max' speed of flywheel}}{\text{motor speed}} = \frac{\text{motor pulley dia'}}{\text{flywheel dia'}}$$

$$\text{Max speed of flywheel} = \frac{0.48\,\text{m} \times 10\,\text{rev/s}}{1.7\,\text{m}}$$

$$= 2.824\,\text{rev/s}$$

$$\text{Radius of gyration } k^2 = \frac{R_1^2 + R_2^2}{2}$$

$$k^2 = \frac{1.7^2 + 1.22^2}{2}$$

$$= 2.189\,\text{m}^2$$

$$\text{At 2.824 rev/s} \times \omega_1 = 2.824 \times 2 \times 2\pi$$

$$= 17.744\,\text{rad/s}$$

$$\text{At (2.824} - 0.25\,\text{rev/s)} = 2.574 \times 2$$

$$= 16.173\,\text{rad/s}$$

$$\text{Moment of inertia } mk^2$$

$$= 1902.21 \times 2.189\,\text{kgm}^2$$

$$= 4163.94\,\text{kgm}^2$$

$$\text{Kinetic energy at 2.824 rev/s} = \frac{1}{2}\omega^2$$

$$= \frac{1}{2} \times 4163.94\,\text{kgm}^2 \times 17.744^2\,\text{rad/s}$$

$$= 0.6555\,\text{MNm}$$

$$\text{Kinetic energy at 2.574 rev/s} = \frac{1}{2} \times \omega_2^2$$

$$= \frac{1}{2} \times 4163.94\,\text{kgm}^2 \times 16.173^2\,\text{rad/s}$$

$$= 0.5446\,\text{MNm}$$

$$\text{Energy given out} = 0.6555 - 0.5446\,\text{MNm}$$

$$= 0.1109\,\text{MNm}$$

$$\text{Energy absorbed or work done at blade} = 0.1109\,\text{MNm} \times 0.8$$

$$= 0.08872\,\text{MNm}$$

Work done = average cutting force × distance moved

$$\text{Average cutting force} = \frac{0.08872 \text{ MNm}}{0.075 \text{ m}} \times 10^3$$

$$= 1182.9 \text{ kN}$$

$$\text{Area of section being sheared} = \frac{\text{force}}{\text{shear stress}}$$

$$= \frac{1182.9 \text{ kN}}{350 \times 10^3 \text{ kN/m}^2}$$

$$= 0.00338 \text{ m}^2$$

$$= 3379.8 \text{ mm}^2$$

12 Gearing

It is not the intention of this chapter to give an in-depth analysis of the design of gear systems. For a more comprehensive study on the design of gearing see *Design Engineer's Handbook* (Chapter 12, Introduction to Geared Systems) or *Dudley's Gear*. Only a few brief examples of the primary form of tooth gearing will be given here; the two forms being considered are:

1. Spur gearing
2. Bevel gearing

12.1 SPUR GEARING

General formula:

$$E = \frac{T}{r} = \frac{Power}{\omega r} = \frac{Power}{V} \tag{12.1}$$

Lewis formula for strength (American Gear Manufacturers Association):

$$\sigma = \frac{W \cos \phi}{K_v . F . m . y} \tag{12.2}$$

where

$$W = \frac{19.1P}{N.D}$$

$$y = 0.484 - \frac{4.24}{t+6}$$

$$K_v = \frac{3.54}{3.54 + \sqrt{V}}$$

$$\phi = \text{pressure angle } (20°)$$

12.1.1 NOTATION

For gear notation see Table 12.1. See Figure 12.1 for a general description of a gear set.

12.1.2 WORKING STRESS σ_w

Allowable values for σ_w are given in Table 12.2. These cover the more common materials used in gear manufacture.

12.1.3 WIDTH OF TEETH

Using the Lewis factor to find the wheel proportions involves fixing the width in terms of the circular pitch and substituting the appropriate formula. Generally for slow speeds and where shafts are inaccurately adjusted, the face width may be 1.25 to 2.5 times the pitch (the average works out

TABLE 12.1

Gear Notation

$E =$	Tangential load on teeth	(N)
$T =$	Torque transmitted	(Nm)
$r =$	Pitch circle radius	(m)
$V =$	Pitch line velocity	(m/s)
$y =$	Lewis form factor	(See notes)
$b =$	Width of teeth	(m)
$P =$	Circular pitch	(m)
$\sigma_w =$	Safe working stress	(MPa)
$D =$	Pitch circle diameter for wheel	(m)
$d =$	Pitch circle diameter for pinion	(m)
$T =$	Number of teeth for wheel	(m)
$t =$	Number of teeth for pinion	(m)
$m =$	Module	(m)
$A =$	Addendum = module	(m)
$B =$	Dedendum = $1.25 \times$ module	(m)

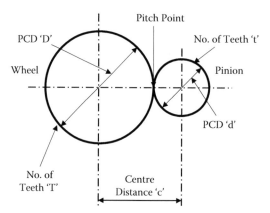

FIGURE 12.1   Nomenclature for spur gear set.

TABLE 12.2

Lewis Factor Stresses (σ_w) in MPa

Material	Pitch Circle Velocity (m/min)							
	30	60	120	180	270	370	550	730
Cast iron	54.4	40.9	32.7	27.2	20.4	16.3	13.9	11.7
Cast steel	136	102	81.7	68.0	51.0	40.8	34.7	29.7
Forged steel	163	122.5	95.3	81.7	68.0	51.0	44.2	37.2
Phosphor bronze	71.5	71.5	44.2	37.4	32.6	27.2	20.8	16.6
Nickel chrome steel (1544 MPa ult.)	—	—	224.6	187.0	150.0	126.0	93.6	75.9

to 3 to 4 times the pitch); for high speeds, smooth engagement and high wear, the width may be 6 to 8 times. High ratios result in finer tooth pitches.

Example 12.1

Two spur gears are to have a ratio of 4:3 and a 6 mm module. If the centre distance is to be approximately 125 mm, calculate:

1. The circular pitch.
2. The number of teeth on each gear.
3. The pitch circle diameters.
4. The exact centre distance.

Solution:

1. The circular pitch:

$$\rho = \pi m \qquad\qquad (12.3)$$

$$= 6\pi$$

$$= 18.849 \text{ mm}$$

2. The number of teeth on each gear:

$$\text{From } 0.5(D + d) = \text{centre distance}$$

$$D + d = 125 \times 2 \qquad\qquad (12.4)$$

Also

$$m = \frac{D}{T} = \frac{d}{t}$$

$$6 = \frac{D}{T} \quad \text{and}$$

$$6T = D \quad \text{and} \quad 6t = d$$

Substituting in Equation (12.2),

$$6T + 6t = 250 \qquad\qquad (12.5)$$

Also T/t = 4/3 and substitute in Equation (12.3) for T.

$$6 \times \frac{4}{3} t + 6t = 250$$

$$\text{hence } t = 17.86$$

Both gears must contain a whole number of teeth: therefore to make T a whole number t must be divisible by 3.

Hence the nearest multiple of 3 is 18.

$$\text{therefore } T = \frac{4}{3} \times 18$$

$$T = 24 \text{ teeth}$$

$$t = 18 \text{ teeth}$$

3. The pitch circle diameters:

$$\text{Wheel pitch circle diameter } D = T \times \text{module} \qquad (12.6)$$

$$= 24 \times 6$$

$$= 144 \text{ mm}$$

$$\text{Pinion pitch circle diameter } d = t \times \text{module} \qquad (12.7)$$

$$= 18 \times 6$$

$$= 108 \text{ mm}$$

4. Exact centre distance:

$$c = 0.5 \times (D + d) \qquad (12.8)$$

$$= 0.5 \times (144 + 108)$$

$$= 126.0 \text{ mm}$$

Example 12.2

Design a pair of involute spur gears to transmit power of 30 kW at 8.5 rev/s. Gear ratio of 4:1. Assume a 20 tooth pinion of 6 mm module and a ratio of face width to pitch of 4. The pinion is forged steel and the wheel is cast steel.

Find the tooth proportions of wheel and pinion.

Solution:

$$\text{Torque} = \frac{\text{power}}{2\pi N}$$

$$= \frac{30 \times 10^3}{2\pi \times 8.5} \frac{\text{Nm}}{\text{s}} \times \text{s}$$

$$= 561.72 \text{ Nm}$$

Also torque = E × r; here E is the tangential load and

$$E \times \frac{\text{pcd}}{2} = 561.72 \text{ Nm}$$

Also

$$\text{pcd} = \frac{t \cdot p}{\pi}$$

$$E \times \frac{tp}{2\pi} = 561.72 \text{ Nm}$$

$$E = \frac{2\pi \times 561.72}{t \times p}$$

Assuming 20° involute, full depth:

$$y = 0.154 - \frac{0.912}{t} \text{ and } t = 20$$

$$y = 0.1084$$

Pitch circle diameter:

$$\text{pcd of pinion (d)} = \text{module} \times t$$

$$= 6 \times 20 \text{ mm}$$

$$= 120 \text{ mm}$$

And circular pitch:

$$= \frac{\pi d}{t} = \frac{\pi \times 120 \text{ mm}}{20} \text{ or } (\pi \times m)$$

$$= 18.85 \text{ mm}$$

Given b = 4 × circular pitch,

$$b = 4 \times 18.85$$

$$= 75.40 \text{ (say 75.0 mm)}$$

To find allowable working stress where

$$\sigma_w = \frac{E}{ybp}$$

$$\sigma_w = \frac{2\pi .562 \text{ Nm}}{t.p^2.y.b}$$

$$\sigma_w = \frac{2\pi \times 562 \text{ Nm}}{20 \times 0.0189^2 \text{m} \times 0.1084 \times 0.075 \text{ m}}$$

$$= 60.796 \text{ MPa}$$

To find pitch line velocity:

$$= \pi d.\text{rev/min}$$

$$= \pi \times 0.12 \text{ m} \times 8.5 \text{ rev/s} \times 60$$

$$= 192.265 \text{ m/min}$$

From Table 12.2 for a forged steel pinion, 180 m/min gives 81.7 MPa and 270 m/min gives 68 MPa.

By interpolation 192.3 m/min is equivalent to 79.83 Mpa.

The calculated stress value σ_w of 60.8 MPa would therefore be satisfactory.

Blank outside diameter for pinion:

$$= d + (2 \times \text{module})$$

$$= 120 \text{ mm} + (2 \times 6 \text{ mm})$$

$$= 132.0 \text{ mm}$$

Tangiental load at pitch line:

$$E = \frac{562 \text{ Nm}}{\text{radius of pinion}}$$

$$= \frac{562 \text{ Nm}}{0.06 \text{ m}}$$

$$= 9366.7 \text{ N}$$

Load per mm of tooth width:

$$= \frac{9366.7 \text{ N}}{75 \text{ mm}}$$

$$= 124.9 \text{ N/mm}$$

Data for Cast Steel Wheel

Gear ratio is 4 to 1 and number of teeth in the pinion is 20. Therefore the number of teeth in the wheel = 20 × 4 = 80.

Blank outside diameter for wheel:

$$= D + 2 \text{ m} \tag{12.9}$$

$$= 4 \times 120 \text{ m} + 2 \times 6 \text{ m}$$

$$= 492 \text{ mm}$$

Stress in wheel:

$$\sigma_w = \frac{E}{ybp} \tag{12.10}$$

$$= \frac{9367\text{N}}{\mathbf{Y} \times 0.075\text{m} \times 0.0189\text{m}}$$

$$\text{where } y = 0.154 - \frac{0.912}{80t}$$

$$= 0.1426$$

$$\sigma_w = \frac{9366N}{0.1426 \times 0.075m \times 0.0189m}$$

$$= 46.335 \text{ MPa}$$

σ_w for cast steel from Table 12.2, 180 m/min gives 68 MPa and 270 m/min yields 51 MPa. By interpolation, 180 m/min is equivalent to 65.73 MPa; therefore cast steel wheel stress is satisfactory.

Wheel and pinion shafts would be sized using the procedure outlined in Example 3.1. It is calculated that a shaft diameter of 66 mm, including keyway, allowance will be acceptable.

Table 12.3 summarises the gear details for this example.

Wheel Arms

Because of the large diameter of the wheel and as it is cast, it would be considered desirable to design the wheel with spokes:

1. To reduce the amount of cast metal in the melt and reduce any shrinkage problems
2. To reduce the mass and therefore the inertia in the wheel

For the wheel arms (or spokes) it is considered that four arms would be reasonable and elliptical in cross section (see Figure 12.2).

$$\text{Boss diameter} = 2 \times 66$$

$$= 132 \text{ mm (2 x bore diameter)}$$

$$\text{Bending arm radius} = \frac{480 \text{ mm}}{2} - \frac{132 \text{ mm}}{2}$$

$$= 174 \text{ mm}$$

and

$$b = 22 \text{ mm}$$

$$\text{Tangential load} = 9366 \text{ N}$$

$$\text{Load per wheel arm} = 2341.5 \text{ N}$$

$$\text{Bending moment} = 2341.5 \text{ N} \times 0.174 \text{ m}$$

$$= 407.42 \text{ Nm}$$

$$\text{Bending moment} = \sigma Z$$

TABLE 12.3
Summarised Particulars for Example 12.2

Item	No. of Teeth	Module (mm)	PC Diameter (mm)	Tooth Width (mm)	Blank Diameter (mm)
Forged steel pinion	20	6	120	75	132
Cast steel wheel	80	6	480	75	492

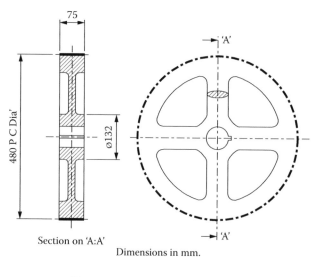

Section on 'A:A'

Dimensions in mm.

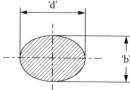

FIGURE 12.2   Detail of spur gear wheel used in Example 12.2.

For elliptical section:

$$Z = \frac{\pi b d^2}{32}$$

Say 3b = d and allowable bending stress for cast steel allowing for load reversal:

$$\sigma = 44 \text{ MPa}$$

$$407.42 \text{ Nm} = 44 \times 10^6 \times \frac{\pi}{32} \times \frac{d}{3} \times d^2$$

$$\frac{407.42 \times 32}{44 \times 10^6 \times \pi} = \frac{d}{3} \times d^2$$

$$94.324 \times 10^{-6} \times 3 = d^3$$

$$d = 65.65 \text{ mm}$$

12.2   BEVEL GEARING

A skeletal layout of a pair of bevel gears is shown in Figure 12.3. The more important features used in the calculation of the gear proportions that are required for the manufacture are identified.

Since the bevel tooth is only a full tooth at the pitch cone diameter, the method for calculating the tooth proportions differs slightly from that used for the ordinary spur wheel and pinion.

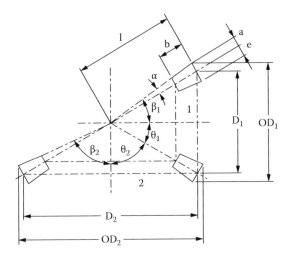

FIGURE 12.3 Nomenclature for bevel gear set.

The following modified form of the Lewis formula has given satisfactory results. The tooth load is assumed to act at the mean radius of the tooth width, and the formula has a correction to allow for this.

12.2.1 MODIFIED LEWIS FORMULA FOR BEVEL GEARS

$$E = r^2 ybp\sigma_w \qquad (12.11)$$

where
   r = ratio of mean pitch radius at pitch radius
   y = form factor based on the number of teeth in the equivalent spur gear
   E = tangential load at pitch circle (N)
   b = width of tooth (m)
   p = circular pitch (m)
   σ_w = allowable working stress (N/m²)

The width of the tooth will usually vary from 1/4 to 1/3, the length of the pitch cone surface.

Note: Where the formula is common to both the pinion and the wheel, the suffix is omitted.

The definitions and basic formulae used in the calculation of bevel gears is covered in Table 12.4.

Example 12.3

Design a pair of bevel gears to transmit 25 kW at 8.5 rev/s of the driving shaft and have a reduction ratio of 2.5:1 with the shafts at right angles to each other. See Figure 12.4 for details.

Solution:

Both gears are to be cast steel. For this problem assume the pinion has 24 teeth and an involute angle of 20°, together with a module of 6 mm.

$$\text{Mean torque} = \frac{\text{power}}{2\pi \times N}$$

TABLE 12.4

Basic Formulae for Bevel Gears

Symbol	Definition	Formula
t	No. of teeth in pinion	$t = \pi d/p$
T	No. of teeth in wheel	$T = \pi D/p$
D	Pitch circle diameter	$D = mT$ or $d = mt$
O.D.	Outside or blank diameter	$OD = D + 2a \cos \theta$
a	Addendum	$a = module$
e	Dedendum	$e = 1.25\ m$
θ_1	Pinion pitch cone angle	$\theta_1 = \tan^{-1} t/T$
θ_2	Wheel pitch cone angle	$\theta_2 = \tan^{-1} T/t$
β	Face angle	$\beta = \theta + \tan^{-1} a/l$
p	Circular pitch	$p = \pi m$
m	Module	$D/T = m = d/t$
n_1	Number of teeth in equivalent spur wheel	$n_1 = t/\cos \theta$
l	Length of pitch cone surface	$l = D/(2 \sin \theta)$

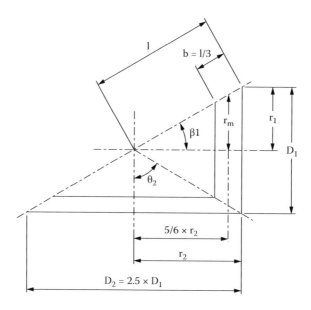

FIGURE 12.4   Bevel gear set used in Example 12.3.

$$= \frac{25 \times 10^3 \text{ Nm}}{2\pi \times 8.5 \text{ rev/s}}$$

$$= 468.10 \text{ Nm}$$

Pitch cone angles:

$$t = \text{number of teeth in pinion}$$

$$= 24 \text{ teeth}$$

$$T = \text{number of teeth in wheel}$$

$$= 24 \times 2.5$$

$$= 60 \text{ teeth}$$

Pinion pitch cone angle:

$$\theta_1 = \tan^{-1} \frac{t}{T}$$

$$= \tan^{-1} \times \frac{24}{60}$$

$$\theta_1 = 21.80°$$

For shafts at right angles:

Wheel pitch cone angle:

$$\theta_2 = 90° - 21.48°$$

$$= 68.20°$$

Note: See Table 12.4 for a summary of the gear sizes

Design of teeth:

$$E = r^2 y b p \sigma_w$$

To find ratio (r), let width of teeth = 1/3 height of pitch cone surface, i.e.:

$$h = 1/3 \times l \text{ (see Figure 12.4)}$$

$$r_1 = \text{pitch radius of pinion}$$

$$r_m = \text{mean pitch radius of pinion}$$

$$\frac{r_m}{r_1} = \frac{5}{6} = r$$

By similar triangles to find b in terms of the circular pitch (p).

$$l = \frac{D_1}{2 \sin \theta_1}$$

$$= \frac{tp}{2\pi} \sin \theta_1$$

where

$$\theta_1 = 21.80°$$

$$t = 24$$

$$b = \frac{tp}{6\pi \sin \theta_1}$$

$$= \frac{24 \times p}{6\pi \times 0.3714}$$

$$= 3.428\, p$$

To find the value of y, the first step is to find the number of teeth in the equivalent spur wheel for pinion:

$$n^1 = \frac{t}{\cos \theta_1}$$

$$= \frac{24}{0.9305}$$

$$= 25.79 \quad (\text{say } \underline{26})$$

For 20° involute full tooth:

$$y = 0.154 - \frac{0.912}{26}$$

$$= 0.1189$$

Now torque = 468.10 Nm and also

$$\text{torque} = E \times \frac{D_1}{2} \quad \text{or} \quad 468.10 = \frac{tp}{2\pi}$$

i.e. tangential load:

$$E = \frac{2\pi \times 468.10}{24P}$$

$$= \frac{122.55}{P}$$

Working stress:

$$\frac{122.55}{P} = r^2 ybp\sigma_w \times \quad \frac{122.55 \times 36}{p^3 \times 3.428 \times 25 \times 0.119} = \sigma_w$$

where

$$p = \pi m$$

$$= \pi \times 6$$

$$= 18.85 \text{ mm}$$

$$= 0.01885 \text{ m}$$

from which:

$$\sigma_w = 64.3 \text{ MPa}$$

This value is compared with Table 12.2 for pitch circle velocity of $\pi n \times d$ where d is found as follows:

For pinion pitch diameter:

$$D_1 = m \times t$$

$$= 6 \text{ mm} \times 24t$$

$$= 144 \text{ mm}$$

For wheel pitch diameter:

$$D_2 = m \times T$$

$$= 6 \text{ mm} \times 60t$$

$$= 360 \text{ mm}$$

Therefore:

$$\text{Pitch line velocity} = \pi \times 510 \text{ rev/min} \times 0.144 \text{ m}$$

$$= 230.719 \text{ m/min (say 231 m/min)}$$

From Table 12.2 for cast steel:

180 m/min gives 68 MPa.
270 m/min gives 51 MPa.

By interpolation, 231 m/min gives:

$$\sigma_w = 60.63 \text{ MPa}$$

Therefore stress by calculation is considered satisfactory.
    To find the width of tooth:

$$= 3.428 \times p$$

$$= 3.428 \times 18.9$$

$$= 64.79 \text{ mm}$$

Blank or outside diameters:

For pinion:

$$DO_1 = D_1 + 2a \cos\theta_1$$

$$= 144 \text{ mm} + 2 \times 6 \times \cos 21.8°$$

$$= 155.142 \text{ mm}$$

For wheel:

$$DO_2 = D_2 + 2a \cos \theta_2$$

$$= 360 \text{ mm} + 2 \times 6 \times \cos 68.2°$$

$$= 364.456 \text{ mm}$$

Face angles:

$$\beta = \theta + \tan^{-1} \frac{a}{l}$$

$$l = \frac{D_1}{2 \sin \theta_1}$$

$$= \frac{144}{2 \times \sin 21.8°}$$

$$= 193.877 \text{ mm}$$

$$\gamma = \tan^{-1} \frac{a}{l}$$

$$= \tan^{-1} \left(\frac{6}{193.877} \right)$$

$$= \tan^{-1} (0.03095)$$

This is common for both pinion and wheel.
For pinion:

$$\text{Face angle } \beta_1 = 21.8° + \tan^{-1} 0.03095$$

$$= 23.57°$$

For wheel:

$$\text{Face angle } \beta_2 = 68.2° + \tan^{-1} 0.03095$$

$$= 69.97°$$

Wheel Arms

The load on the arms is calculated from the torque at the mean pitch radius, and the tangential tooth load at the centre of the teeth would be:

$$E = \frac{\text{Mean torque}}{\text{Mean pitch radius}}$$

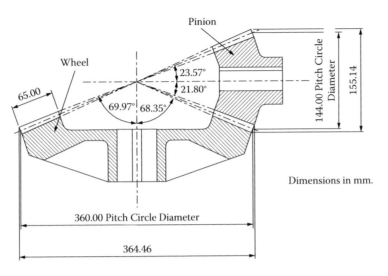

FIGURE 12.5 Final details for bevel gears in Example 12.3.

TABLE 12.5
Summary of Gear Details for Example 12.3

Item	No. of Teeth	Module (mm)	PC Diameter	Width of Teeth	Diameter of Blank	Pitch Cone Angle	Face Angle
Pinion	24	6	144	64.8	155.14	21.8°	23.57°
Wheel	60	6	360	64.8	364.46	68.2°	69.97°

From the example above:

$$E = \frac{468\,\text{Nm}}{(5/6)\ \times\ 0.144\,\text{m}}$$

$$= 3900\,\text{N}$$

The arms would then be calculated to resist this load.
The final details of the bevel gear pair are as shown in Figure 12.5 and Table 12.5.

13 Introduction to Material Selection

13.1 INTRODUCTION

One of the more difficult tasks encountered in engineering design is deciding the most appropriate material to use. With a wide choice of steels to choose from, it may seem daunting to a young engineer when faced with the decision to make the correct choice.

When a design is being modified or updated, the selection of material may have already been decided and so the problem will not arise. There may be a case where the material normally used has become obsolete and the material supplier has ceased to supply it. They may suggest a choice, and it will be up to the designer to select the most appropriate one. The design may have had a minimal modification, say including a change in cross section, and then the original choice can be retained. But where there has been a substantial change in the component section due to a redesign to, say, reduce weight, then the engineer may have to consider carrying out a series of stress calculations to ensure that the life of the component is not compromised.

These notes are proposed as an aid for the young engineer when studying the important issues in material selection. Table 13.1 summarises the material characteristics and properties to be considered for new and critical design applications.

The first objective in selecting the correct material is to carry out a simple stress analysis as outlined in Chapter 1; this will then determine the approximate stress levels the component is being subject to. If the maximum stress is found to be in the order of, say, 450 MPa, then there is no point in selecting a material such as, say, aluminium, which has an ultimate strength of, for instance, 450 MPa. In this case there would be no Factor of Safety (FoS) if the loading were to be exceeded. A better choice would be a medium carbon steel with an ultimate strength of 620 MPa giving a minimum FoS of 1.3.

13.2 THINGS TO CONSIDER

The following is a basic checklist that the student engineer will need to address as the design progresses. Some or all of the information may not be immediately available in the early phases of the design and will need to be acquired as the design phase progresses. It is important that the essential information is available before the initial design review.

13.2.1 ENVIRONMENT

The nature of the environment to which the component or structure will be exposed will need to be established at the earliest possible stage. If the atmosphere is in an engineering workshop at ambient temperature, then most corrodible materials will be suitable, but if the environment is in an offshore situation, then the choices will be limited to high strength corrosion resisting steels. If the component will be used in the aerospace environment, then obviously the choice will be limited to high strength lightweight materials such as aluminium, but as with the offshore industry, care will need to be taken to select a suitable corrosion protection package.

TABLE 13.1

Summary of Material Characteristics and Properties to Be Considered for Critical Design Applications

1.1	**Static characteristics**		**1.4**	**Thermal properties**
	Strength			Coefficient of linear expansion
	Ultimate strength			Thermal shock resistance
	Yield strength		**1.5**	**Manufacturability**
	Shear strength			Productability
	Density			Availability
	Ductility			Machinability
	Young's modulus			Weldability
	Poisson's ratio			Heat treatment
	Hardness			Formability
	Form			Spinning
	Sectional dimensions (ruling section)			Deforming (forging)
1.2	**Fatigue characteristics**		**1.6**	**Corrosion environment**
	Life required			Seawater
	Vibration and shock			Galvanic corrosion
	Fatigue strength			Stress corrosion
	Spectrum loads			
	Corrosion fatigue			
1.3	**Fracture characteristics**			
	Fracture toughness			
	Flaw growth			
	Crack stability			

13.2.2 STRENGTH

There are three principal usages of strength:

- Static strength: The ability to resist short term steady loads at ambient temperature
- Dynamic strength: The ability to resist a fluctuating load
- Creep strength: The ability to resist a load at elevated temperatures to produce a progressive extension over an extended period of time

13.2.3 DURABILITY

The dictionary definition of *durability* is "the ability to exist a long time without any significant deterioration." In the context of this chapter this will include resistance to wear and abrasion and corrosion attacks.

13.2.4 STIFFNESS

Stiffness is the ability of a material to maintain its shape when subject to a load or force. Consider Hooke's law, where a test material is incrementally loaded to produce an extension which is plotted against the load and the resultant slope is used to demonstrate the relationship between stress and strain. Within the linear range of the extension, the material will return to its original size when the load or force is removed. It the test material is loaded so that this linear extension is reached, the material is said to have reached its *proportional limit*; any further extension past this limit will then result in the test material not returning back to its original size when the applied load or force is removed. This phase is called the non-proportional limit.

A material that displays a steep stress-strain curve will be stiffer than one that has a shallow curve, and a component manufactured from the first material will deflect less for a given load or force than one manufactured from the second.

13.2.5   WEIGHT

In the transport industry, including road and aerospace vehicles, it is paramount to keep weight to a minimum. In both cases fuel is required to propel the vehicle, and with the current high cost of fuel it is essential to maximise the distance travelled per unit of fuel used. Substantial efforts are made in the design of lightweight structures.

The opposite is true in the case of a machine tool, where weight is important to minimise the effects of vibration and structural distortion when subject to cutting or forming forces.

13.2.6   MANUFACTURING

In the current atmosphere of high material costs efforts are being made by manufacturers to minimise material wastage. In the 19th and early to mid-20th centuries it was not uncommon to generate significant amounts of wastage from metal cutting machines; in the current financial climate efforts are being made to minimise this wastage by making more use of forging, casting and other metal forming techniques, such as pressure die casting when the material properties allow its use. In the automotive and aerospace industries significant use is made of fabrications manufactured from formed sections.

13.2.7   COST

In any manufacturing enterprise, regardless of size, every effort is made to minimise cost. If the product manufactured is destined for the consumer market, then the selling price will be dictated by the buying public and the manufacturing cost will have to be recovered from this; the difference will be the profit generated. When the manufacturer is selling through retailers, then the retailers will demand a percentage of the selling price, resulting in less profit for the manufacturer. The manufacturer will have to be very cost conscious and scrutinise where cost savings can be made in the product manufacture.

13.2.8   MAINTAINABILITY

A manufacturer has to make a decision whether to consider product maintenance. In the case of a low cost product such as a hairdryer it may be argued that this will be a throw-away item at the end of the product's life and not to consider any maintenance. On the other hand, in a high capital cost item, such as a road vehicle, maintenance becomes a critical issue and efforts are made to ensure the product is maintainable throughout its life span with the use of replaceable modules that can be returned either to the manufacturer or companies specialising in repairing and offering the items back for replacement.

13.3   A MODEL FOR MATERIAL SELECTION

The author has developed a material selection model similar to the one produced by Stuart Pugh and Bill Hollins for "total design." Figure 13.1 shows the basic model, and it is intended to put the decision making into a chronological order so that the important issues are dealt with in a timely manner and not considered out of order. A brief description of the main headings follows.

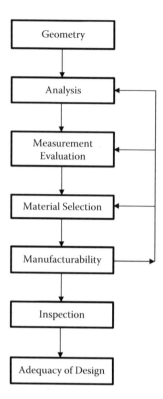

FIGURE 13.1   Model for material selection (Core activities).

13.3.1 Geometry

The shape of the component is identified and all the boundary conditions defined. External and internal forces can then be evaluated.

13.3.2 Analysis

A detailed stress analysis will need to be carried out next to determine the stresses generated by the applied forces. In the initial stages of the design this will be to try and establish the maximum and minimum stresses to enable a preliminary selection to be made. This activity may be revisited several times as better definition becomes available.

13.3.3 Measurement Evaluation

Where possible, a suitable prototype is constructed and tested to confirm the measured stresses and deflections agree with the calculated values. The model can be either a full size physical model or a finite element model if there is sufficient confidence in the model's accuracy.

13.3.4 Material Selection

The stage has been reached where a suitable candidate material can now be considered. Each industry will have its own specific requirements; for example, the aerospace and automotive industry will have strength and density constraints as a high priority, whereas the agriculture industry will require cost and resistance to corrosion as their priorities.

13.3.5 Manufacturability

The manufacture of the component will have an important influence on the final selection of the material depending upon whether the component is machined, fabricated, cast or forged. The production department will usually reach this decision depending upon the range of machines at its disposal. The designer may have to revisit the design and make any suitable changes to the geometry and review the stress analysis.

13.3.6 Adequacy of Design

In keeping with the current total design philosophy, it is important that the final design is carefully reviewed to ensure that it will meet the original design specification and that any changes made do not compromise the design integrity.

13.4 A MATERIAL DATABASE

There are a number of sources for material data, notably MIL-HDBK 5, which is published by the U.S. government, and also British Standards in the UK publishes a range of standards covering a wide range of materials.

The majority of manufacturing companies retain a stock of materials to support their manufacturing activity, and these tend to cover the most common materials used.

If a new material is used that is not in stock or is not a stock item at the local material suppliers, then a financial penalty may have to be paid if the material is unusual and has to be manufactured outside the normal range of sizes, etc. This will apply to rolled or extruded sections where special tooling has to be designed and manufactured. This will be an important consideration in the design of a suitable company based database where only readily supplied material and sizes are included.

Databases can be held in two different forms:

- Paper based
- Computer based

13.4.1 Paper Based Database

As outlined above, paper databases are available from government or commercial agencies. These will be comprehensive and cover a very wide range of materials not required at the company level. They cover the majority of material properties and are a very useful source of data. One disadvantage with this type of database is that they are soon out of date, and therefore need to be re-issued periodically to include new materials and delete those materials that have become obsolete.

13.4.2 Computer Based Database

A computer based database can be available on line from a subscription service; the advantage is that it is updated as new materials are added. Like the paper based database, it includes materials that the user does not necessarily require.

The young engineer should produce a personal database of material properties for his or her own personal future reference. Figure 13.2 gives details of the template for database input used by the author for metallic materials. The template has space for more data than the reader may require, but gives room for expansion, as not every engineer will require the same type of information. The template can be modified to suit other purposes. The data generated can then be input into a suitable database program such as Microsoft® Access or Excel, where a search facility exists, allowing for suitable candidate materials to be searched and extracted.

Template for Database Input

                                           Material Specification No.

Specification			
Form		—	
Family		—	
Aircraft specification		—	
General Properties			
Atomic number		—	
Atomic weight		—	
Density	(ρ)	—	Mg/m^3
Basic alloying elements		—	
Price	(P)	—	£/kg
Mechanical Properties			
Ultimate strength	(f_t)	—	Mpa
Yield strength (0.2% proof)	(t_2)	—	Mpa
Yield strength (0.1% proof)	(t_1)	—	Mpa
Compressive stress (0.2% proof)	(c_2)		Mpa
(0.1% proof)	(c_1)		Mpa
Shear strength	(f_{so})	—	Mpa
Ultimate bearing stress	(f_b)		Mpa
Bearing stress	(b_{10})	—	Mpa
Ultimate torsional stress	(F_q)	—	Mpa
Torsional proof stress (0.1% proof)	(Q_1)	—	Mpa
Poisson's ratio	(v)	—	
Elongation	$(5.65\sqrt{So})$	—	% longitudinal
			% transverse
Ductility	(εf)	—	
Toughness	(K_{Ic})	—	MPA $m^{1/2}$
Hardness	(H)	—	HV
Modulus of elasticity	(E)	—	Gpa
Shear modulus	(G)	—	Gpa
Bulk modulus		—	

FIGURE 13.2 Template for database input.

Modulus of rupture		—	Mpa	
Fatigue ratio		(R)	—	
Impact resistance	(Izod)	—	J longitudinal	
			J transverse	
Loss coefficient	(η)	—		
Thermal Properties				
Melting point	(Tm)	—	°C	
Glass temperature	(Tg)	—	°C	
Minimum service temperature	(Tmin)	—	°C	
Maximum service temperature	(Tmax)	—	°C	
Specific heat	(Cp)	—	kJ/kg°C	
Thermal expansion		(α)	—	$10^{-6}/°C$ (20–200°C)
Thermal shock reserve		—		
Latent heat of fusion		—		
Thermal conductivity	(λ)	—	W/m°Cat20°C	
Weldability				
Electron beam		Resistance to corrosion		
Arc		Resistance to stress corrosion cracking		
Resistance (overlap joint)		Resistance to exfoliation corrosion		
Flame		Magnetic properties		
Brazing				
Minimum welding properties				
Typical Uses				
Remarks				
Main References				

FIGURE 13.2 (*Continued*)

13.4.3 MATERIAL CLASSIFICATION AND CODING

A suitable classification and coding scheme needs to be developed to enable the data to be stored within the database. The author uses the following description in the construction of his own database. Tables 13.2 and 13.3 suggest a possible classification and coding scheme for use in the current exercise. Table 13.2 lists the possible material groups. Table 13.3 lists the sub-group numbers for the primary forms.

Table 13.4 shows a section of the material group table with the respective sub-group numbers against the material types. The respective database tables correlate the group and sub-group numbers together with their respective properties. For example, heat resisting alloy forging would

TABLE 13.2
Performance Indices for Minimum Weight

Component Shape and Loading	Stress Limited	Stiffness Limited
Rods in tension	σ_{tu}/ρ	E/ρ
Short columns in compression	σ_{tu}/ρ	E/ρ
Thin wall pipes and pressure vessels under internal pressure	σ_{tu}/ρ	—
Helical springs for a specified length and load capacity	σ_s/ρ	—
Thin wall shafts in torsion	σ_s/ρ	G/ρ
Rods and pins in shear	σ_s/ρ	—
Beams with a fixed section shape, in bending	$\sigma_{tu}^{2/3}/\rho$	$E^{1/2}/\rho$
Solid shafts in torsion	$\sigma_s^{2/3}/\rho$	$G^{1/2}/\rho$
Solid shafts in bending	$\sigma_{tu}^{2/3}/\rho$	$E^{1/2}/\rho$
Rectangular beams with fixed width, in bending	$\sigma_{tu}^{1/2}/\rho$	$E^{1/3}/\rho$
Flat plates under pressure	$\sigma_{tu}^{1/2}/\rho$	$E^{1/3}/\rho$

Notation

Density	ρ
Modulus of elasticity	E
Shear modulus	G
Ultimate tensile strength	σ_{tu}
Ultimate shear stress	σ_s

come under the Group 5000 and the Sub-Group 170 for forging; therefore the material code number will be 5170. MS is prefixed to the code number defining it as a material specification. The database would then be constructed using the listings outlined in Table 13.5. As this is based on using a relational database, the respective database tables are then related as per the table. Tables 13.6, 13.7, 13.8, 13.9, 13.10 and 13.11 are used.

Each table is to be populated with the respective details of the materials included in the database. As new materials are added, the database will become more comprehensive and useful.

When all the tables have been completed, it is then a simple matter to create a database report to identify candidate materials that meet the required specified criteria.

Table 13.12 is an example of a partly completed table that is covering material indices. Table 13.13 is a partly completed table of material properties that is used in the solution of the following two examples, which show how the database is used to identify candidate materials that meet a specified criterion.

Example 13.1

Consider a cantilever with a single static load applied at its free end.

From a previous calculation it was established that the factored stress in the section was 400 MPa due to the load and an appropriate Safety Factor (SF). There is a further requirement that the component be as light as possible to minimise weight.

The designer puts together a simple list of criteria for the material to initially meet, from which a limited range of candidate materials was identified.

The list comprises:

1. Density (from database in Table 13.1): To be less than 2.8 kg/M³.
2. Ultimate stress (from database in Table 13.2): To be greater than 400 MPa.
3. Material performance index: From Table 3.2 stress limited design for rectangular beam in bending. $\sigma_{tu}^{1/2}/\rho$ to be greater than 15.

TABLE 13.3
Performance Indices for Elastic Design

Component and Design Requirement	Maximise
Springs Specifies energy storage, volume to be minimised.	σ_{tu}^2/E
Springs Specifies energy storage, mass to be minimised.	$\sigma_{tu}^2/E\rho$
Compression seals and gaskets Maximum contact area with specified maximum contact pressure.	σ_{tu}/E
Diaphragms Maximum deflection under specified pressure or force.	$\sigma_{tu}^{3/2}/E$
Ties, columns Maximum longitudinal vibration frequencies.	E/ρ
Beams Maximum longitudinal vibration frequencies.	$E^{1/2}/\rho$
Plates Maximum flexural vibration frequencies.	$E^{1/3}/\rho$
Ties, columns, beams, plates Maximum self-damping.	η

Notation

Density	ρ
Modulus of elasticity	E
Ultimate tensile strength	σ_{tu}
Loss coefficient	η

The results of a search based upon the above criteria are shown in Table 13.14.

From the above search, five candidate materials are identified as meeting the material requirements.

Maximising the material performance index (σ_{tu}/ρ) defines MS 13201 as being the better performer. MS 13201 (L168) is supplied in the bar and extruded section, making it the ideal choice for manufacture.

Example 13.2

Consider an internally pressurised cylinder.

The requirement is for a thin walled cylinder of 150 mm internal diameter with a minimum wall thickness and an internal pressure of 13.8 MPa (2000 lbf/in.[2]). The material should exhibit a good ductility so that the component will not fail due to brittle fracture. Weight is not a problem.

The search criteria will be:

1. Ultimate stress (from database in Table 13.7): To be greater than 500 MPa.
2. Elongation (from database in Table 13.7): To be greater than 25%.
3. Material performance index (from database in Table 13.8): From Table 3.2 stress limited design for thin wall. Pipes and pressure vessels under internal pressure. σ/ρ to be greater than 65.0.

TABLE 13.4
Part Proposed Classification System for Metallic Materials

Group	Material	Sub-Group	Form
100000	Aluminium alloys		
	11000	Low strength	
		111000	Wire
		112000	Bar
		113000	Sheet and strip
		114000	Plate
		115000	Tube
		116000	Casting
		117000	Forging
	12000	Medium strength	
		121000	Wire
		122000	Bar
		123000	Sheet and strip
		124000	Plate
		125000	Tube
		126000	Casting
		127000	Forging
	13000	High strength	
		131000	Wire
		132000	Bar
		133000	Sheet and strip
		134000	Plate
		135000	Tube
		136000	Casting
		137000	Forging
200000	Magnesium alloys		
	21000	Low strength	

TABLE 13.5
Relationship between Database Tables

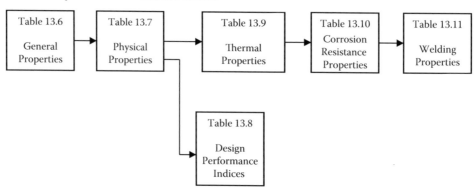

TABLE 13.6
General Properties

1	Atomic no.
2	Atomic weight
3	Density
4	Basic alloying elements
5	Magnetic properties
6	Price per unit volume

TABLE 13.7
Physical Properties

1	Ultimate strength	(f_{tu})
2	Yield strength	(t_2)
3	Yield strength	(t_1)
4	Compressive strength	(c_2)
5	Compressive strength	(c_1)
6	Shear strength	(f_{so})
7	Ultimate bearing strength	(f_b)
8	Bearing stress	(b_{10})
9	Ultimate torsion stress	(F_q)
10	Torsional proof stress (0.1%)	(Q_1)
11	Poisson's ratio	(v)
12	Elongation	$(5.65\sqrt{So}$
13	Ductility	(εF)
14	Toughness	(K_{1c})
15	Hardness	(H)
16	Modulus of elasticity	(E)
17	Shear modulus	(G)
18	Fatigue ratio	(R)
19	Impact resistance	(Charpy)
20	Impact resistance	(Izod)

TABLE 13.8
Design Performance Indices

1	E/ρ
2	$E^{1/2}/\rho$
3	$E^{1/3}/\rho$
4	$G^{1/2}/\rho$
5	σ/ρ
6	$\sigma^{2/3}/\rho$
7	$\sigma^{1/2}/\rho$
8	σ/E
9	σ^2/E
10	$\sigma^2/E\rho$
11	$\sigma^3/E2$
12	$\sigma^{2/3}/E$

TABLE 13.9
Thermal Properties

1	Melting point	(T_m)
2	Glass temperature	(T_g)
3	Minimum service temperature	(T_{min})
4	Maximum service temperature	(T_{max})
5	Specific heat	(C_p)
6	Coefficient of thermal expansion ($\times 10^{-6}$)	(α)
7	Thermal shock reserve	
8	Latent heat of fusion	
9	Thermal conductivity	(λ)

TABLE 13.10
Corrosion Resistance Properties

1	Resistance to corrosion
2	Resistance to stress corrosion cracking
3	Resistance to exfoliation cracking

TABLE 13.11
Welding Properties

1	Flame
2	Arc
3	Resistance
4	Brazing
5	Electron beam

The result of the search based upon the above criteria is shown in Table 13.15.

MS 33002 is BS HR203, plate, sheet and strip. This obviously would be unsuitable to manufacture a pressure vessel from. The remaining material, MS 52002, is BS S129, a corrosion resisting steel, and this is now obsolete. MS 52003 is BS S130, corrosion resisting steel, and both are supplied in the form of bar and forgings.

MS 52003 would be the logical choice for this application. The elongation is better than specified, and the performance index is also very good. It will be easy to calculate the required minimum wall thickness using a factored ultimate stress in the thin wall stress calculation pd/2t.

13.5   FUTURE DEVELOPMENTS

13.5.1   Knowledge Based Engineering (KBE)

It became obvious in the early 1990s that a parametric language could be developed for use in a CAD system where the operator would simply fill in a table and the system would then generate the appropriate model. Amendments to the design could then be implemented by changes to the variables within the table. This approach will be able to save many man-hours where the design is modular.

TABLE 13.12
Part Table of Material Indices

Material	E $\times 10^9$	G $\times 10^9$	ρ $\times 10^0$	σ $\times 10^6$	σ/ρ $\times 10^6$	$\sigma^{1/2}/\rho$ $\times 10^6$	$\sigma^{2/3}/\rho$ $\times 10^6$	E/ρ $\times 10^9$	$E^{1/3}/\rho$ $\times 10^9$	$E^{1/2}/\rho$ $\times 10^9$	σ/E $\times 10^{-3}$	σ^2/E $\times 10^{-3}$	σ^3/E^2 $\times 10^{-3}$	$\sigma^{3/2}/E$ $\times 10^{15}$	$\sigma^2/E\rho$ $\times 10^{-3}$	$G^{1/2}/\rho$ $\times 10^0$
MS 11501	72	27	2.69	160	59.480	4.702	10.956	26.766	1.547	3.154	2.222	355.556	790.123	28.109	132.177	1.932
MS 11601	72	27	2.69	230	85.502	5.638	13.955	26.766	1.547	3.154	3.194	734.722	2347.029	48.446	273.131	1.932
MS 11602	71	27	2.71	215	79.336	5.411	13.243	26.199	1.528	3.109	3.028	651.056	1971.509	44.402	240.242	1.917
MS 12201	70	27	2.7	310	114.815	6.521	16.965	25.926	1.526	3.099	4.429	1372.857	6079.796	77.973	508.466	1.925
MS 12301	70	27	2.7	295	109.259	6.361	16.413	25.926	1.526	3.099	4.214	1243.214	5239.260	72.383	460.450	1.925
MS 12401	70	27	2.7	295	109.259	6.361	16.413	25.926	1.526	3.099	4.214	1243.214	5239.260	72.383	460.450	1.925
MS 12501	70	27	2.7	310	114.815	6.521	16.965	25.926	1.526	3.099	4.429	1372.857	6079.796	77.973	508.466	1.925
MS 12701	70	27	2.7	295	109.259	6.361	16.413	25.926	1.526	3.099	4.214	1243.214	5239.260	72.383	460.450	1.925
MS 13201	72	27	2.8	490	175.000	7.906	22.198	25.714	1.486	3.030	6.806	3334.722	22,694.637	150.647	1190.972	1.856
MS 13301	72	27	2.7	440	162.963	7.769	21.426	26.667	1.541	3.143	6.111	2688.889	16,432.099	128.188	995.885	1.925
MS 13401	72	27	2.8	450	160.714	7.576	20.973	25.714	1.486	3.030	6.250	2812.500	17,578.125	132.583	1004.464	1.856
MS 13501	72	27	2.8	400	142.857	7.143	19.389	25.714	1.486	3.030	5.556	2222.222	12,345.679	111.111	793.651	1.856
MS 13701	72	27	2.8	450	160.714	7.576	20.973	25.714	1.486	3.030	6.250	2812.500	17,578.125	132.583	1004.464	1.856
MS 32001	183	85	8.19	1000	122.100	3.861	12.210	22.344	0.693	1.652	5.464	5464.481	29,860.551	172.802	667.214	1.126
MS 32002	193	75	8.18	1080	132.029	4.018	12.869	23.594	0.706	1.698	5.596	6043.523	33,818.680	183.899	738.817	1.059
MS 33001	214		8.18	1080	132.029	4.018	12.869	26.161	0.731	1.788	5.047	5450.467	27,507.031	165.852	666.316	
MS 33002	210	80	8.37	650	77.658	3.046	8.965	25.090	0.710	1.731	3.095	2011.905	6227.324	78.913	240.371	1.069
MS 52001	193	77	7.72	880	113.990	3.843	11.895	25.000	0.749	1.800	4.560	4012.435	18,295.041	135.259	519.745	1.137
MS 52002	193	76	7.9	540	68.354	2.942	8.394	24.430	0.732	1.759	2.798	1510.881	4227.335	65.018	191.251	1.104
MS 52003	193	76	7.92	540	68.182	2.934	8.373	24.369	0.730	1.754	2.798	1510.881	4227.335	65.018	190.768	1.101
MS 52004	211	80	7.83	930	118.774	3.895	12.168	26.948	0.760	1.855	4.408	4099.052	18,066.912	134.413	523.506	1.142
MS 52005	211		7.82	1130	144.501	4.299	13.873	26.982	0.761	1.858	5.355	6051.659	32,409.357	180.026	773.869	
MS 52006	201	77	7.82	1270	162.404	4.557	14.997	25.703	0.749	1.813	6.318	8024.378	50,701.295	225.169	1026.135	1.122
MS 53001	170		7.9	772	97.722	3.517	10.652	21.519	0.701	1.650	4.541	3505.788	15,920.403	126.176	443.771	
MS 53002	170		7.92	772	97.475	3.508	10.626	21.465	0.699	1.646	4.541	3505.788	15,920.403	126.176	442.650	
MS 53003	193		7.92	540	68.182	2.934	8.373	24.369	0.730	1.754	2.798	1510.881	4227.335	65.018	190.768	
MS 54001	193	76	7.92	550	69.444	2.961	8.476	24.369	0.730	1.754	2.850	1567.358	4466.563	66.832	197.899	1.101

(Continued)

TABLE 13.12 (Continued)
Part Table of Material Indices

Material	E $\times 10^9$	G $\times 10^9$	ρ $\times 10^0$	σ $\times 10^6$	σ/ρ $\times 10^6$	$\sigma^{1/2}/\rho$ $\times 10^6$	$\sigma^{2/3}/\rho$ $\times 10^6$	E/ρ $\times 10^9$	$E^{1/3}/\rho$ $\times 10^9$	$E^{1/2}/\rho$ $\times 10^9$	σ/E $\times 10^{-3}$	σ^2/E $\times 10^{-3}$	σ^3/E^2 $\times 10^{-3}$	$\sigma^{3/2}/E$ $\times 10^{15}$	$\sigma^2/E.\rho$ $\times 10^{-3}$	$G^{1/2}/\rho$ $\times 10^0$
MS 56001	193	78	7.7	950	123.377	4.003	12.550	25.065	0.751	1.804	4.922	4676.166	23,017.396	151.715	607.294	1.147
MS 56002	193	78	7.7	1200	155.844	4.499	14.665	25.065	0.751	1.804	6.218	7461.140	46,390.507	215.385	968.979	1.147
MS 56003	200	77	7.92	460	58.081	2.708	7.524	25.253	0.738	1.786	2.300	1058.000	2433.400	49.330	133.586	1.108
MS 62001	176	69	7.85	1320	168.153	4.628	15.329	22.420	0.714	1.690	7.500	9900.000	74,250.000	272.489	1261.146	1.058
MS 62002	204	80	7.84	930	118.622	3.890	12.153	26.020	0.751	1.822	4.559	4239.706	19,328.071	139.025	540.779	1.141
MS 62003	200	77	7.86	1080	137.405	4.181	13.392	25.445	0.744	1.799	5.400	5832.000	31,492.800	177.462	741.985	1.116
MS 62004	204	81	7.86	880	111.959	3.774	11.683	25.954	0.749	1.817	4.314	3796.078	16,375.240	127.966	482.962	1.145
MS 72001	110	42	4.42	900	203.620	6.787	21.090	24.887	1.084	2.373	8.182	7363.636	60,247.934	245.455	1665.981	1.466
MS 73001	106	45	4.51	390	86.475	4.379	11.836	23.503	1.049	2.283	3.679	1434.906	5279.370	72.659	318.161	1.487
MS 77001	110	42	4.42	900	203.620	6.787	21.090	24.887	1.084	2.373	8.182	7363.636	60,247.934	245.455	1665.981	1.466

TABLE 13.13

Part Table of Material Properties

Material	Form	Family	Ultimate Strength f_t MPa	Yield Strength 0.2% proof MPa	Yield Strength 0.1% proof MPa	Compressive Strength 0.2% proof MPa	Compressive Strength 0.1% proof MPa	Shear Strength MPa	Poisson's Ratio	Density Mg/m³	Elongation % Longitude
MS 11501	Tube	Aluminium	160–200	60					0.33	2.69	18
MS 11601	Casting	Aluminium	230	185	175			180	0.33	2.68	2
MS 11602	Casting	Aluminium	215	190	170			172	0.33	2.71	1
MS 12201	Bar	Aluminium	310	270	263			172	0.33	2.7	8
MS 12301	Sheet/strip	Aluminium	295	240	232				0.33	2.7	8
MS 12401	Plate	Aluminium	295	240	231			171	0.33	2.7	8
MS 12501	Tube	Aluminium	310	240	229				0.33	2.7	9
MS 12701	Forging	Aluminium	295	255	248			177	0.33	2.7	8
MS 13201	Bar	Aluminium	490	440	432	391	403	260	0.33	2.8	7
MS 13301	Sheet/strip	Aluminium	440	380	367				0.33	2.8	8
MS 13401	Sheet/strip	Aluminium	460	410	402	421	435	270	0.33	2.8	8
MS 13501	Tube	Aluminium	400	290	275			270	0.33	2.8	10
MS 13701	Forging	Aluminium	450	395	375			270	0.33	2.8	6
MS 32001	Bar	Heat resisting steel	1000	600				800	0.29	8.19	20
MS 32002	Bar	Heat resisting steel	1080	695					0.29	8.18	20
MS 33001	Sheet/strip	Heat resisting steel	1080	695					0.29	8.18	20
MS 33002	Plate	Heat resisting steel	620	230					0.29	8.37	30
MS 52001	Bar	Corrosion resisting steel	880–1080	650	649			810	0.3	7.72	12
MS 52002	Bar	Corrosion resisting steel	540	210	200			417	0.29	7.9	35
MS 52003	Bar	Corrosion resisting steel	540	210	200			417	0.29	7.92	35
MS 52004	Bar	Corrosion resisting steel	930–1080	780	741			614	0.3	7.83	15
MS 52005	Bar	Corrosion resisting steel	1130–1330	1030				746	0.3	7.82	12

(Continued)

TABLE 13.13 (Continued)
Part Table of Material Properties

Material	Form	Family	Ultimate Strength f_t MPa	Yield Strength 0.2% proof MPa	Yield Strength 0.1% proof MPa	Compressive Strength 0.2% proof MPa	Compressive Strength 0.1% proof MPa	Shear Strength MPa	Poisson's Ratio	Density Mg/m³	Elongation % Longitude
MS 52006	Bar	Corrosion resisting steel	1270–1470	1030				838	0.3	7.82	10
MS 53001	Sheet/strip	Corrosion resisting steel	772	636	550	419	361		0.3	7.9	13
MS 53002	Sheet/strip	Corrosion resisting steel	772	636	550	419	361		0.3	7.92	13
MS 53003	Sheet/strip	Corrosion resisting steel	540	210	200				0.29	7.92	35
MS 54001	Tube	Corrosion resisting steel	550–700	210–340					0.29	7.92	
MS 56001	Casting	Corrosion resisting steel	950–1200	800					0.29	7.7	12
MS 56002	Casting	Corrosion resisting steel	1200–1500	960					0.3	7.7	8
MS 56003	Casting	Corrosion resisting steel	460	200					0.3	7.92	
MS 62001	Bar	Non-corrosion resisting steel	1320–1520	1030				803	0.3	7.85	8
MS 62002	Bar	Non-corrosion resisting steel	930–1080	740	695			614	0.3	7.84	13
MS 62003	Bar	Non-corrosion resisting steel	1080–1280	880				713	0.3	7.86	10
MS 62004	Bar	Non-corrosion resisting steel	880–1080	690				581	0.3	7.86	12
MS 72001	Bar	Titanium alloy	900–1160	830	806		869	495	0.31	4.42	8
MS 73001	Sheet/strip	Titanium alloy	390–540	290	274				0.31	4.51	22
MS 77001	Forging	Titanium alloy	900–1160	830	806			495	0.31	4.42	8

TABLE 13.14
Candidate Materials to Satisfy the Search
Criteria for Cantilever: Example 13.1

Material Spec	Density kg/m³	Ultimate Strength MPa	$\sigma_{tu}^{1/2}/\rho$ $\times 10^6$
MS13201	2.80	490	7.9057
MS13301	2.80	440	7.7689
MS13401	2.80	450	7.5761
MS13501	2.80	400	7.1429
MS13701	2.80	450	7.5761

TABLE 13.15
Candidate Materials for the Pressure Vessel:
Example 13.2

Material Spec	Ultimate Strength MPa	Elongation %	σ/ρ $\times 10^6$
MS 33002	650	30	77.6583
MS 52002	540	35	68.3544
MS 52003	540	35	68.1818

A number of researchers have developed knowledge based systems that could be integrated into the design process, allowing domain specific knowledge to be stored regarding a part or process, together with other associated attributes.

Further work has been extended to develop the knowledge base using a material database together with a suitable solid modelling system that uses a rule based technique, for example, using an "if-then" approach which is implemented to perform the material selection process. A material that satisfies all the constraints then becomes the most suitable candidate for a particular component operating within a set of specific conditions.

14 General Tables

Tables 14.1, 14.2 and 14.3 give allowable stress values for a steady or permanent working load covering a range of common structural materials.

TABLE 14.1
Using a Safety Factor of 5—Allowable Working Stress for a Steady or Permanent Working Load

Material	Type of Stress			
	Tension (t) (MPa)	Compression (c) (MPa)	Bending (b) (MPa)	Shear (MPa)
Cast iron	29	83	41–55	27.5
Mild steel	89.5–117	89.5–117	89.5–117	69–117
Cast steel	117–145	117–145	117–145	89.5–117
Steel casting	55–69	83–110	69–96.5	16.5
Rolled copper	41.5	—	—	—
Brass	21	—	—	—

TABLE 14.2
Allowable Working Stress for a Load Varying from Zero to Maximum Value

Material	Tension (t) (MPa)	Compression (c) (MPa)	Bending (b) (MPa)	Shear (MPa)
Cast iron	19.3	58.6	27.5	24
Mild steel	59.3–78.6	59.3–83	59.3–78.6	45–59.3
Cast steel	78.6–96.5	78.6–96.5	78.6–96.5	59.3–78.6
Steel casting	36.5–55	55–73	45.5–65	32.4–55
Rolled copper	20.7	—	—	11.0
Brass	13.8	—	—	—

TABLE 14.3
Allowable Working Stresses for a Fluctuating Load Producing Equal Stresses in Opposite Directions

Material	Tension and Compression (MPa)	Bending (MPa)	Shear (MPa)
Cast iron	9.7	13.8	12.0
Mild steel	29.6–39.6	29.6–39.4	22.8–29.6
Cast steel	39.3–48.3	39.3–48.3	29.6–39.3
Steel casting	18.6–27.6	22.8–32.4	15.9–27.6

TABLE 14.4
Specific Gravity of Engineering Materials

Material	Specific Gravity
Aluminium—cast	2.56
Aluminium—wrought	2.67
Brass—cast	8.10
Brass—sheet 75% Cu	8.45
Copper—cast	8.79
Copper—sheet	8.81
Copper—wire	8.91
Gunmetal 83% Cu	8.46
Gunmetal 79% Cu	8.73
Iron—cast	6.90–7.50
Iron—wrought (average)	7.70
Lead	11.40
Monel	8.87
Phosphor bronze—cast	8.60
Steel	7.73–7.90
Tin—hammered	7.39
Tin—pure	7.29
Zinc—cast	6.86
Zinc—rolled	7.19

TABLE 14.5
Basic SI Units

Quantity	Name of Unit	Symbol	Definition
Mass	Kilogram	kg	
Length	Metre	m	
Time	Second	s	
Energy	Joule	J	$= Nm = kg\ m^2/s^2$
Power	Watt	W	$= J/s = kg\ m^2/s^3$
Pressure	Pascal	Pa	$= N/m^2 = kg/m.s^2$
Force	Newton	N	$= m.a = kg.m/s^2$
Frequency	Hertz	Hz	$= s^{-1}$

Note: a = acceleration.

TABLE 14.6
ISO Metric Coarse Threads

Outside Diameter	Core Diameter	Pitch	Depth	Effective Diameter	Tapping Drill	Clearance Diameter
1.6	1.1706	0.35	0.2147	1.373	1.25	1.65
1.8	1.3706	0.35	0.2147	1.573	1.45	1.85
2.0	1.5092	0.40	0.2454	1.740	1.60	2.05
2.2	1.6480	0.45	0.2760	1.908	1.75	2.25
2.5	1.9480	0.45	0.2760	2.208	2.05	2.60
3.0	2.3866	0.50	0.3067	2.675	2.50	3.10
3.5	2.7638	0.60	0.3681	3.110	2.90	3.60
4.0	3.1412	0.70	0.4294	3.545	3.30	4.10
4.5	3.5798	0.75	0.4601	4.013	3.80	4.60
5.0	4.0184	0.80	0.4908	4.480	4.20	5.10
6.0	4.7732	1.00	0.6134	5.350	5.00	6.10
7.0	5.7732	1.00	0.6134	6.350	6.00	7.20
8.0	6.4664	1.25	0.7668	7.188	6.80	8.20
10.0	8.1596	1.50	0.9202	9.026	8.50	10.20
12.0	9.8530	1.75	1.0735	10.863	10.20	12.20
14.0	11.5462	2.00	1.2269	12.701	12.00	14.25
16.0	13.5462	2.00	1.2269	14.701	14.00	16.25
18.0	14.9328	2.50	1.5336	16.376	15.50	18.25
20.0	16.9328	2.50	1.5336	18.376	17.50	20.25
22.0	18.9328	2.50	1.5336	20.376	19.50	22.25
24.0	20.3194	3.00	1.8403	22.051	21.00	24.25
27.0	23.3194	3.00	1.8403	25.051	24.00	27.25
30.0	25.7060	3.50	2.1470	27.727	26.50	30.50
33.0	28.7060	3.50	2.1470	30.727	29.50	33.50
36.0	31.0924	4.00	2.4538	33.402	32.00	36.50
39.0	34.0924	4.00	2.4538	36.402	35.00	39.50
42.0	36.4790	4.50	2.7605	39.077	37.50	42.50
45.0	39.4790	4.50	2.7605	42.077	40.50	45.50
48.0	41.8646	5.00	3.0672	44.752	43.00	48.75
52.0	45.8646	5.00	3.0672	48.752	47.00	52.75
56.0	49.2522	5.50	3.3739	52.428	50.50	56.75
60.0	53.2522	5.50	3.3739	56.428	54.50	60.75
64.0	56.6388	6.00	3.6806	60.103	58.00	64.75
68.0	60.6388	6.00	3.6806	64.103	62.00	68.75

Note: All dimensions in mm.

TABLE 14.7
General Units

Dimensions:

Mass = M, length = L, time = T.

Heat energy:

Unit of heat energy is the Joule (J).

Mechanical equivalent of heat energy = 4.186 J/calorie

1 calorie (cal) = quantity of heat required to raise 1 g of water through 1°C

Angular measure:

1 radian = 360/(2π) = 57.296°

TABLE 14.8
Greek Alphabet

Alpha	A	a	Nu	N	ν
Beta	B	β	Xi	Ξ	ξ
Gamma	Γ	γ	Omicron	O	o
Delta	Δ	δ	Pi	Π	π
Epsilon	E	ε	Rho	P	ρ
Zeta	Z	ζ	Sigma	Σ	σ
Eta	H	η	Tau	T	τ
Theta	Θ	θ	Upsilon	Y	υ
Iota	I	ι	Phi	Φ	φ
Kappa	K	κ	Chi	X	χ
Lambda	Λ	λ	Psi	Ψ	ψ
Mu	M	μ	Omega	Ω	ω

TABLE 14.9
Mathematical Expressions

y ∞ x:	y is proportional to x.
y = f(x):	y is a function of x.
y ≈ x:	y is approximately equal to x.
y ≠ x:	y is not equal to x.

TABLE 14.10
Other Important Units

Force:

One Newton (1 N) is the force required to give a mass of one kilogram (1 kg) an acceleration of one metre per second per second.

$$1 \text{ N} = 1 \text{ kg} \times \text{m/s}^2$$

Energy:

One Joule (1 J) is the work done by a force of one Newton (1 N) when moving through a distance of one metre (1 m).

$$1 \text{ J} = 1 \text{ N} \times 1 \text{ m}$$

Pressure:

One Pascal (1 P) is the pressure exerted by a force of one Newton acting on an area of one square metre (1 m^2).

$$1 \text{ Pa} = 1 \text{ N/m}^2$$

Power:

One watt (1 W) is the power due to a force of one Newton moving at a speed of one metre per second or due to a torque of one Newton metre (Nm) when rotating at a speed of one radian per second.

$$1 \text{ W} = 1 \text{ N} \times 1 \text{ m/s}$$

or

$$1 \text{ W} = 1 \text{ Nm} \times 1 \text{ rad/s}$$

TABLE 14.11
Galvanic Corrosion

Galvanic Series in flowing sea water.		
Alloy	Voltage Range of Alloy vs. Reference Voltage[1]	
Anodic or Active End		**Corroded**
Magnesium	−1.60 to −1.63	
Zinc	−0.98 to −1.03	
Aluminium Alloys	−0.70 to −1.90	
Cadmium	−0.70 to −0.76	
Cast Irons	−0.60 to −0.72	
Steel	−0.60 to −0.70	
Aluminium Bronze	−0.30 to −0.40	
Naval Brass	−0.30 to −0.40	
Copper	−0.28 to −0.36	
Lead-Tin Solder (50/50)	−0.26 to −0.35	
Admiralty Brass	−0.25 to −0.34	
Manganese Bronze	−0.25 to −0.33	
Silicon Bronze	−0.24 to −0.27	**Direction**
Stainless Steel (400 Series)[2]	−0.20 to −0.35	**of Attack**
Copper Nickel (90-10)	−0.21 to −0.28	
Lead	−0.19 to −0.25	
Copper Nickel (70-30)	−0.13 to −0.22	
Stainless Steel (17-4 PH)[3]	−0.10 to −0.20	
Silver	−0.09 to −0.14	
Monel	−0.04 to −0.14	
Stainless Steel (300 Series)[2][3]	0.00 to −0.15	
Titanium and Titanium Alloys [3]	+0.06 to −0.05	
Inconel 625[3]	+0.10 to −0.04	
Hastelloy C-276[3]	+0.10 to −0.04	
Platinum[3]	+0.25 to +0.18	
Graphite	+0.30 to +0.20	
Cathodic or Noble End		**Protected**

[1] These numbers refer to saturated calomel electrodes.

[2] In poorly aerated or low velocity water, or inside crevices, these alloys may start to corrode and exhibit potentials closer to −0.5 V.

[3] When covered with slime films of marine bacteria, these alloys may exhibit potentials between +0.3 and +0.4 V.

Bibliography

The following book titles, although not all referenced directly, will be useful to students of design.

Constrato, *Steel Designer's Manual*, 4th ed., Oxford, UK: Wiley-Blackwell, 1989.

Richards. K.L. *Design Engineer's Handbook*, Boca Raton, FL: CRC Press, 2012.

Timoshenko. S. and Goodier J. N. *Theory of Elasticity*, 2nd ed., New York: McGraw-Hill, 1951.

Pilkey, W.D. and Pilkey, D.F. *Peterson's Stress Concentration Factors*, 2nd ed., Wiley-Interscience 2007.

Pugh, S. *Total Design: Integrated Methods for Successful Product Engineering*, Wokingham UK: Addison-Wesley, 1990.

Radzevich, S.P. (ed.) *Dudley's Handbook of Practical Gear Design and Manufacture.* 2nd ed., Boca Raton, FL: CRC Press, 2012.

U.S. Department of Defense, MIL-HDBK-5J, Metallic Materials and Elements for Aerospace Vehicle Structures, 2003.

Young. W.C. *Roark's Formulas for Stress and Strain*, 6th ed., New York: McGraw-Hill, 1989.

Index

A

Attachments, methods of, 69–78
  attachment types, 69
  bolts in tension, 69–74
    angle bracket, 71
    bolts in shear due to eccentric loading, 73–74
    bracket subject to eccentric loading, 73
    centre of rotation, 69
    detail of angle bracket, 70
    direct load, 72
    example, 69, 71, 73
    heeled bracket, 69
    heel point, 71
    loading producing a tensile load in bolt, 69–71
    load producing a tension and shear load in bolt, 71–73
    maximum stresses, 73
    maximum tension in bolts, 70
    permissible stress, 71, 73
    solution, 69, 72, 73
    tensile load, 72
    top fastener, resultant stress on, 72
    turning moment, load on fastener due to, 74
  welding (permanent), 75–78
    allowable working stress, 77
    eccentric turning effect, 76
    example, 75
    polar second moment of area of the weld arrangement, 76
    resultant vector, 77
    sectional properties of weld, 76
    solution, 76
    strength of welded joints, 75–78
    stress concentration factors of various welds, 75
    throat thickness, 78
    welded bracket, 75
    weld throat dimensions, 75

B

Beam sections subject to bending, 19–29
  basic theory, 19–21
  beam classification, 19
  columns, 19
  common sectional properties, 24
  description of beam, 19
  parallel axis theorem, 21–29
    beam section, 27
    bending moment diagram, 22
    example, 22, 23, 27, 29
    Factor of Safety, 24
    identification of beam sections, 28
    individual and combined inertia calculations, 26
    loaded beam, 22
    maximum bending moment, 23

    maximum compressive stress in top flange, 26
    maximum deflection, 27
    maximum tensile strength in bottom flange, 26
    physical properties of sections, 28
    position of neutral axis, 26
    revised diameter, 23
    Safety Factor, 22
    second moment of area of a section, 21
    section beam, 25
    solution, 22, 23, 27
    spreadsheet program, 25
    ultimate tensile stress, 23, 27
  standard bending cases, 20
  standard loading cases for horizontal beams, 20–21
  struts, 19
Bevel gearing, 122–129
  basic formulae for bevel gears, 124
  bevel gear set, 124
  circular pitch, 125
  example, 123
  fine details for bevel gears, 129
  full tooth, 122
  modified Lewis formula for bevel gears, 123–129
  pinion pitch cone angle, 125
  pinion pitch diameter, 127
  pitch cone angles, 124
  skeletal layout, 122, 123
  solution, 123
  summary of gear details, 129
  tangential load, 126
  tooth load, 123
  wheel arms, 128
  wheel pitch cone angle, 125
  wheel pitch diameter, 127
  working stress, 126
Bolts in tension, 69–74
  angle bracket, 71
  bolts in shear due to eccentric loading, 73–74
  bracket subject to eccentric loading, 73
  centre of rotation, 69
  detail of angle bracket, 70
  direct load, 72
  example, 69, 71, 73
  heeled bracket, 69
  heel point, 71
  loading producing a tensile load in bolt, 69–71
  load producing a tension and shear load in bolt, 71–73
  maximum stresses, 73
  maximum tension in bolts, 70
  permissible stress, 71, 73
  solution, 69, 72, 73
  tensile load, 72
  top fastener, resultant stress on, 72
  turning moment, load on fastener due to, 74
Bulk modulus, 12

C

Clevis and Clevis pin, 8
Columns and struts, 79–85
  background, 79–80
    empirical formulae, 79
    end fixing conditions, 80
    Euler's theory, 79
    slenderness ratio, definition of, 79
    Young's modulus, 79
  Perry-Robertson method, 84–85
    behaviour of practical columns, 84
    critical value for stress, 84
    Euler's predictions and, 84, 85
    notes on use, 84
  Rankine-Gordon method, 80–84
    buckling effects, 81
    crane gantry, column for, 82
    Euler's formula, 81
    example, 81, 82
    Factor of Safety, 84
    formula, 80
    long columns, 81
    mild steel, 83
    safe load, 80
    short columns, 81
    solution, 81, 82
    universal beam, 82
Compressive stress, 1
Creep strength, 132
Cylinders, *see* Thick cylinders

D

d'Alembert's principle, 53
Direct strains, 2–3
  definition of strain, 2
  example, 2
  solution, 2–3
  symbol for strain, 2
Direct stress, 1
Distortion-energy theory, 17
Domain specific knowledge, 147
Double shear, 8
  Clevis and Clevis pin, 8
  example, 8
  pinned lug, 8
  solution, 8
Dynamic strength, 132

E

Eccentric loading, 73–74, 87–90
  enlarged section across section a:a, 89
  example, 87
  general formulae, 87
  maximum compressive stress, 89
  maximum tensile stress, 89
  moments of inertia about axis x:x, 90
  outline elevation of frame, 88
  principle of superposition, 88
  radius of gyration, 87, 89
  situation, 87
Elastic failure, theories of, 14–17

complex stress theory, 14
direct stress in tensile test, 14
failed ductile specimen, 14
modulus of elasticity, definition of, 14
Rankine's principal stress theory, 16
  example, 16
  safety factor, 16
  solution, 16
shear strain energy theory (Von Mises theory), 17
  distortion-energy theory, 17
  example, 17
  Mohr's circle, 17
  solution, 17
  Von Mises–Hencky criterion for ductile failure, 17
  Von Mises stress, 17
St. Venant's maximum principal strain theory, 16
  example, 16
  maximum principal strain theory, 16
  solution, 16
tensile test failure, 15
typical stress–strain curve, 15
volumetric strain, 15
Energy formulae (flywheel basics), 105–113
  average cutting force, 113
  basic equations for flywheel systems, 106
  coefficient of speed fluctuation, 105
  energy absorbed, 112
  example, 107, 110, 111
  flywheel uses, 105
  kinetic energy, 105, 112
  motor generator, 110
  permissible rim velocity to generate stress, 108
  simple flywheel, 106
  solution, 108, 111
  speed fluctuations, 108
  speed of rotation, 106
  tangential and radial stresses in a flywheel, 107
  theory, 107
  total moment of inertia, 110
Euler's theory, 79

F

Factor of Safety (FoS), 16
  combined torsion and bending, 54
  keys and spline calculations, 63
  parallel axis theorem, 24
  Rankine-Gordon method, 84
  Rankine's principal stress, 16
  shear mode, 14
  straight sided spline calculation, 64
  tensile or compressive mode, 13
    brittle materials, 13
    conditions, 13
    ductile materials, 13
    example, 13
    solution, 13
    yield stress, 13
  thick cylinders, 104
"Fit for function" shaft, 31
Flat plates, 91–98
  bending moment, 92, 98
  bolt diameter, 95

circular flat plates (constant thickness), 92
cover thickness, 91
edge constraint conditions, 91
end plate, 95
example, 91, 95
flange thickness, 95
flat plate cover, 94
load on end cover, 96
loading arrangement, 97
part section of cover, 94
rectangular plate, 91, 93
semi-circle of joint, 97
solution, 91, 96
theory, 91
wall thickness, 96
Flywheels basics (energy formulae), 105–113
average cutting force, 113
basic equations for flywheel systems, 106
coefficient of speed fluctuation, 105
energy absorbed, 112
example, 107, 110, 111
flywheel uses, 105
kinetic energy, 105, 112
motor generator, 110
permissible rim velocity to generate stress, 108
simple flywheel, 106
solution, 108, 111
speed fluctuations, 108
speed of rotation, 106
tangential and radial stresses in a flywheel, 107
theory, 107
total moment of inertia, 110
FoS, see Factor of Safety
Free-body diagram, 42

G

Galvanic corrosion, 154
Gearing, 115–129
bevel gearing, 122–129
basic formulae for bevel gears, 124
bevel gear set, 124
circular pitch, 125
example, 123
fine details for bevel gears, 129
full tooth, 122
modified Lewis formula for bevel gears, 123–129
pinion pitch cone angle, 125
pinion pitch diameter, 127
pitch cone angles, 124
skeletal layout, 122, 123
solution, 123
summary of gear details, 129
tangential load, 126
tooth load, 123
wheel arms, 128
wheel pitch cone angle, 125
wheel pitch diameter, 127
working stress, 126
forms, 115
spur gearing, 115–122
allowable working stress, 119
bending arm radius, 121
circular pitch, 115, 117, 119

data for cast steel wheel, 120
example, 117, 118
finer tooth pitches, 117
general formula, 115
Lewis factor stresses, 116
Lewis formula for strength, 115
notation, 115, 116
pinion pitch circle diameter, 118
pitch circle diameter, 119
pitch line velocity, 119
solution, 117, 118
spur gear wheel, 122
stress in heel, 120
tangential load, 118, 120
wheel arms, 121
wheel pitch circle diameter, 118
width of teeth, 115–122
working stress, 115
General tables, 149–154
allowable working stresses for a fluctuating load
producing equal stresses in opposite
directions, 149
allowable working stress for a load varying from zero
to maximum value, 149
basic SI units, 150
galvanic corrosion, 154
general units, 152
Greek alphabet, 152
ISO metric coarse threads, 151
mathematical expressions, 152
other important units, 153
specific gravity of engineering materials, 150
using a safety factor of 5 (allowable working stress for
a steady or permanent working load), 149
Guest's criterion, 46

H

Hooke's law, 3, 132

I

ISO metric coarse threads, 151
ISO straight sided spline, 60

K

KBE, see Knowledge based engineering
Keys and spline calculations, 57–68
connection types, 58
example calculations, 61–68
compressive stress in spline and shaft, 64
compressive stress in spline teeth, 66
Factor of Safety, 63, 65
factors used, 62
input data, 62
input data for involute spline calculation, 65
input data for straight sided spline calculation, 64
involute spline calculations, 65–68
key calculations, 63–64
shaft calculations, 63
shear stress in shaft, 63
shear stress in spline teeth, 65
showing the fatigue life factors for splines, 67

showing keyway/spline design factor, 66
showing spline application factors, 67
showing the spline distribution factors, 67
showing wear life factor for splines, 68
straight spline calculations, 64–65
summary of key, straight sided and involute spline
strengths from examples, 68
feather key, 57
gear tooth, 57
involute spline, 57
pitch errors, 57
procedure for estimating the strength capacity of shaft,
57–58
hollow shaft, 58
nomenclature for key, 59
nomenclature for key and spline, 58
nomenclature for straight sided spline, 59
reduced diameter for keys and splines, 58
solid shaft, 58
straight spline, 57
strength capacity of ISO involute spline, 60–61
compressive stress in teeth, 61
nomenclature for involute spline, 61
pitch errors, 61
shear stress, 61
strength capacity of ISO straight sided spline, 60
strength capacity of key, 58–60
factors, 59–60
nomenclature for key, 60
values of factors, 60, 66, 67, 68
Kinetic energy (flywheels), 105, 112
Knowledge based engineering (KBE), 142–147
amendments to design, 142
domain specific knowledge, 147
"if-then" approach, 147
rule based technique, 147

L

Lamés theory, 99
Layshaft, 32
Lewis formula for strength, 115
"Line shafting," 32

M

Material database, 135–142
candidate materials for the pressure vessel, 147
candidate materials to satisfy the search criteria for
cantilever, 147
computer based database, 135
corrosion resistance properties, 142
database forms, 135
design performance indices, 141
example, 138, 139
general properties, 141
material classification and coding, 137–142
paper based database, 135
part proposed classification system for metallic
materials, 140
part table of material indices, 143–144
part table of material properties, 145–146
performance indices for elastic design, 139
performance indices for minimum weight, 138

physical properties, 141
relationship between database tables, 140
safety factor, 138
sources for material data, 135
stock of materials, 135
template for database input, 136–137
thermal properties, 142
welding properties, 142
Material selection, introduction to, 131–147
considerations, 131–133
cost, 133
creep strength, 132
durability, 132
dynamic strength, 132
environment, 131
Hooke's law, 132
maintainability, 133
manufacturing, 133
material characteristics and properties to be
considered, 132
proportional limit, 132
static strength, 132
stiffness, 132–133
strength, 132
weight, 133
future developments, 142–147
amendments to design, 142
domain specific knowledge, 147
"if-then" approach, 147
knowledge based engineering, 142–147
rule based technique, 147
material database, 135–142
candidate materials for the pressure vessel, 147
candidate materials to satisfy the search criteria for
cantilever, 147
computer based database, 135
corrosion resistance properties, 142
database forms, 135
design performance indices, 141
example, 138, 139
general properties, 141
material classification and coding, 137–142
paper based database, 135
part proposed classification system for metallic
materials, 140
part table of material indices, 143–144
part table of material properties, 145–146
performance indices for elastic design, 139
performance indices for minimum weight, 138
physical properties, 141
relationship between database tables, 140
safety factor, 138
sources for material data, 135
stock of materials, 135
template for database input, 136–137
thermal properties, 142
welding properties, 142
model for material selection, 133–135
adequacy of design, 135
analysis, 134
geometry, 134
manufacturability, 135
material selection, 134
measurement evaluation, 134

objective, 131
  selection of material, 131
Maximum-shear-stress theory, 46
Microsoft Excel®, 25
Modulus of elasticity, 3–4
  definition of, 14
  force and deformation, 3
  Hooke's law, 3
  modulus of elasticity, 4
  plotting stress against strain, 4
  relationship between stress and strain, 3
  symbol, 4
Modulus of rigidity, 6–7
  constant, 7
  gradient of line, 6
  plot of relationship, 6, 7
  section modulus, 36
Mohr's circle, 17

N

Nomenclature
  bevel gear set, 123
  involute spline, 61
  key, 59, 60
  key and spline, 58
  spur gear set, 116
  straight sided spline, 59
Non-proportional limit, 133

O

Overhead gear, 53
Overturning forces acting on fasteners, 52

P

Parallel axis theorem, 21–29
  beam section, 27
  bending moment diagram, 22
  example, 22, 23, 27, 29
  Factor of Safety, 24
  identification of beam sections, 28
  individual and combined inertia calculations, 26
  loaded beam, 22
  maximum bending moment, 23
  maximum compressive stress in top flange, 26
  maximum deflection, 27
  maximum tensile strength in bottom flange, 26
  physical properties of sections, 28
  position of neutral axis, 26
  revised diameter, 23
  Safety Factor, 22
  second moment of area of a section, 21
  section beam, 25
  solution, 22, 23, 27
  spreadsheet program, 25
  ultimate tensile stress, 23, 27
Perry-Robertson method, 84–85
  behaviour of practical columns, 84
  critical value for stress, 84
  Euler's predictions and, 84, 85
  notes on use, 84
Pinion shaft, 48
Poisson's ratio, 9–10

direct force, 10
  example, 10
  solution, 10
  x and y directions, 9
  y direction, 9
Proportional limit, 132

R

Rankine-Gordon method (columns and struts), 80–84
  buckling effects, 81
  crane gantry, column for, 82
  Euler's formula, 81
  example, 81, 82
  Factor of Safety, 84
  formula, 80
  long columns, 81
  mild steel, 83
  safe load, 80
  short columns, 81
  solution, 81, 82
  universal beam, 82
Rankine's principal stress theory, 16
  example, 16
  safety factor, 16
  solution, 16
Rankine's theory, 46

S

Shaft design basics, 31–44
  procedure for design and analysis of a shaft, 31–35
    angular velocity, 35
    basic shaft under torsion, 35
    bearing placement, 32
    bending moment, 32, 33, 34
    calculate the bending moments and shear forces
        acting on the shaft, 32–345
    calculate the critical diameters for the shaft, 35
    calculate the forces acting on the shaft, 32
    design requirements for the shaft, 31
    determine the torsional profile of the shaft, 34–35
    "fit for function" shaft, 31
    gearbox, 31
    geometry of the shaft, 31
    hydraulic pump, 34
    "line shafting," 32
    questions, 35
    road map for design of shaft, 33
    Safety Factor, 35
    typical layshaft, 32
  rotating shafts, 31
  section modulus, 36–44
    angle of twist, 37
    ASME shaft equations, 38– 39
    bending moment diagram, 42
    coefficients for bending stress concentration
        factors, 1
    coefficients for elastic stress concentration factors, 1
    coefficients for evaluating rectangular sections, 44
    comparison between bearing and undercut
        profiles, 42
    critical points, 43
    example, 36, 37

fillet radii and stress concentrations, 39–40
fillets on a shaft, 40
free-body diagram, 42
hollow sections, 36
mechanical properties for a range of common
   materials, 44
modulus of rigidity, 36
properties for some common sections, 43
reliability factors, 38
solid circular sections, 36
solution, 36, 37
square sections, 36
standard torsion equation, 37
stress concentration, symbol for, 40
undercuts, 40–44
stationary shafts, 31
Shear strain, 6
definition of, 6
energy theory, 17
symbol, 6
Shear stress, 4–6
beam subject to transverse force, 5
block of rubber subject to sideways force, 6
definition of, 4
direction of shear, 5
examples, 4
material being punched, 5
pin subject to shear force, 5
shear force, 4
theory of shear process, 5
unit, 4
SI units, 150
Slenderness ratio, definition of, 79
Spline calculations, *see* Keys and spline calculations
Spreadsheet program, 25
Spur gearing, 115–122
general formula, 115
Lewis factor stresses, 116
Lewis formula for strength, 115
notation, 115, 116
width of teeth, 115–122
   allowable working stress, 119
   bending arm radius, 121
   circular pitch, 115, 117, 119
   data for cast steel wheel, 120
   example, 117, 118
   finer tooth pitches, 117
   pinion pitch circle diameter, 118
   pitch circle diameter, 119
   pitch line velocity, 119
   solution, 117, 118
   spur gear wheel, 122
   stress in heel, 120
   tangential load, 118, 120
   wheel arms, 121
   wheel pitch circle diameter, 118
working stress, 115
Static strength, 132
Stress and strain, introduction to, 1–17
bulk modulus, 12
compressive stress, 1
   area calculation, 1
   fundamental unit of stress, 1
converting between stresses and strains, 10–11

direct strains, 2–3
definition of strain, 2
example, 2
solution, 2–3
symbol for strain, 2
direct stress, 1
deduced stress, 1
stressed material, 1
double shear, 8
Clevis and Clevis pin, 8
example, 8
pinned lug, 8
solution, 8
Factor of Safety in shear mode, 14
Factor of Safety in tensile or compressive mode, 13
brittle materials, 13
conditions, 13
ductile materials, 13
example, 13
solution, 13
yield stress, 13
modulus of elasticity, 3–4
force and deformation, 3
Hooke's law, 3
modulus of elasticity, 4
plotting stress against strain, 4
relationship between stress and strain, 3
symbol, 4
modulus of rigidity, 6–7
constant, 7
gradient of line, 6
plot of relationship, 6, 7
Poisson's ratio, 9–10
direct force, 10
example, 10
solution, 10
x and y directions, 9
y direction, 9
relationship between elastic constants, 12–13
bulk modulus, 12
relationship between shear modulus and other
   elastic constants, 13
volumetric strain, 12
shear strain, 6
definition of, 6
symbol, 6
shear stress, 4–6
beam subject to transverse force, 5
block of rubber subject to sideways force, 6
definition of, 4
direction of shear, 5
examples, 4
material being punched, 5
pin subject to shear force, 5
shear force, 4
theory of shear process, 5
unit, 4
tensile stress, 1
symbol for strain, 1
symbol for stress, 1
theories of elastic failure, 14–17
complex stress theory, 14
direct stress in tensile test, 14
failed ductile specimen, 14

modulus of elasticity, definition of, 14
Rankine's principal stress theory, 16
shear strain energy theory (Von Mises theory), 17
St. Venant's maximum principal strain theory, 16
tensile test failure, 15
typical stress–strain curve, 15
volumetric strain, 15
three dimensional stress and strain, 11
ultimate shear stress, 7
    example, 7
    permanent deformation, 7
    solution, 7
    symbol, 7
ultimate tensile stress, 4
    catastrophic break, 4
    example, 4
    failure values, 4
    solution, 4
volumetric strain, 11–12
Struts, *see* Columns and struts
St. Venant's maximum principal strain theory, 16
    example, 16
    maximum principal strain theory, 16
    solution, 16

T

Tables, *see* General tables
Tensile stress, 1
Thick cylinders, 99–104
    assumption, 99
    cast steel cylinder, 100
    cylinder subject to external pressure, 100
    cylinder subject to internal pressure, 99, 100
    example, 100, 101
    Factor of Safety, 104
    gripping force, 103
    Lamés theory, 99
    maximum hoop stress, 101
    safety factor, 103
    slippage torque, 104
    solid shaft, hoop stress for, 102
    solution, 100, 101
    tangential force, 103
    universal coupling, 101
Torsion and bending, combined, 45–56
    bending of shaft, maximum bending moment, 55
    bolts subject to shear load, 51
    brittle materials, 46
    calculation of bending moment, 45
    cast steel bracket, 49
    d'Alembert's principle, 53
    ductile materials, 46
    evaluation of twisting moment, 45
    example, 45, 47, 49, 53
    factors of safety for various applications, 54
    fitted bolts, 51
    Guest's criterion, 46
    keyway, 49, 55
    material selected, 45
    maximum-shear-stress theory, 46
    notation, 46
    overhead gear, 53
    overturning forces acting on fasteners, 52

pinion shaft, 48
Rankine's theory, 46
reactions at fasteners, 52
shear stress on bolt, 51
shock and fatigue factors, 53
solution, 47, 49, 53
spur reduction gear, 47
steel bracket, 50
stress acting on bolts, 51, 52
tangential tooth load, 47
tension in cable supporting descending balance mass, 54
Tresca criterion, 46
wheel shaft, 48
Tresca criterion, 46

U

Ultimate shear stress, 7
    example, 7
    permanent deformation, 7
    solution, 7
    symbol, 7
Ultimate tensile stress (UTS), 4
    catastrophic break, 4
    example, 4
    failure values, 4
    parallel axis theorem, 23, 27
    solution, 4
UTS, *see* Ultimate tensile stress

V

Volumetric strain, 11–12
    change in volume, 11
    hydrostatically strained cube, 11
Von Mises theory, 17
    distortion-energy theory, 17
    example, 17
    Mohr's circle, 17
    solution, 17
    stress, 17
    Von Mises–Hencky criterion for ductile failure, 17

W

Welding (permanent), 75–78
    allowable working stress, 77
    eccentric turning effect, 76
    example, 75
    polar second moment of area of the weld
        arrangement, 76
    resultant vector, 77
    sectional properties of weld, 76
    solution, 76
    strength of welded joints, 75–78
    stress concentration factors of various welds, 75
    throat thickness, 78
    welded bracket, 75
    weld throat dimensions, 75
Wheel shaft, 48

Y

Yield stress, 13
Young's modulus, 79